SpringerBriefs in Molecular Science

Green Chemistry for Sustainability

Series Editor

Sanjay K. Sharma

For further volumes:
http://www.springer.com/series/10045

Giusy Lofrano
Editor

Green Technologies for Wastewater Treatment

Energy Recovery and Emerging Compounds Removal

 Springer

Giusy Lofrano
Department of Civil Engineering
University of Salerno
via ponte don Melillo
84084 Fisciano
Salerno
Italy

ISSN 2191-5407 e-ISSN 2191-5415
ISBN 978-94-007-1429-8 e-ISBN 978-94-007-1430-4
DOI 10.1007/978-94-007-1430-4
Springer Dordrecht Heidelberg New York London

Library of Congress Control Number: 2012934575

Printed on acid-free paper

Springer is part of Springer Science+Business Media (www.springer.com)

Foreword

At a time when the world's population has reached seven billion people, sustainable design and environmental protection are critical to ensuring that water resources will be available for future generations. It is well recognized that there is an energy/water nexus. It takes water to generate energy and energy to treat water. There is great opportunity to make wastewater treatment plants net energy users and even producers since there is 2–4 times the amount of energy embedded in wastewater than it takes to treat it. As we design wastewater treatment plants, it is important to consider the kinds of treatment that will allow us to recover energy. It is also important to recover nutrients for use as fertilizers and to reclaim water for irrigation, since there is also a water/food nexus. Technology exists which allows wastewater to be treated to a level which removes micro-contaminants such as endocrine disruptors and pharmaceuticals which not only impact receiving waters and their uses, but also limits the ability for direct and indirect water reuse to ensure adequate supply of water. The editor, my friend and colleague Dr. Giusy Lofrano, in the framework of this book not only discusses the problems and issues associated with wastewater treatment but offers technologically sound solutions. This book is an asset to all water professionals so they can become knowledgeable in the issues and develop sustainable design for wastewater treatment plants.

Jeanette A. Brown

Preface

For as long as history has been recorded man has polluted and overexploited the environment in the pursuit of his own well-being. However, as everyone knows, nature itself is a source of pollution (volcanic eruption, natural burning, etc.)

Engineers occupy a strategic place in society because of their dualistic role. If on the one hand their work has sometimes a negative impact on the environment, on the other hand they help reducing or even eliminating pollution by developing treatment processes for water, solid waste and air.

According to the Royal Academy of Engineers, Engineering is "the application of scientific principles to the optimal conversion of natural resources into structures, machines, products, systems and processes for the benefit of humankind".

Nowadays, it is well understood that we must make every possible effort to protect the environment. And now more than ever engineers must provide insights leading towards a sustainable standard of living to protect human and environmental health. However, although the concept of sustainable development has been defined more than 20 years ago, to date its application to actual technological context has not been so straightforward.

The main challenge that green engineering has had to face so far is the operational quest for sustainability. This means looking for strategies where water management and energy use are evaluated jointly. In order to meet this challenge, a number of considerations have to be integrated in the design of wastewater treatment plants, e.g., higher levels of removal efficiency of contaminants, and energy and nutrient recovery. To facilitate direct water reuse, research programs for wastewater treatment must be directed towards technologies that require less non-renewable energy sources, reduce the use of hazardous chemicals, and remove contaminants.

Several Life Cycle Assessment studies of wastewater treatment systems have been evaluating competing technologies and consistently identifying the strong influence that energy consumption has on the overall environmental impact. Not to mention the strong influence that also sludge handling and disposal process have on the overall environmental impact.

In light of these aspects of wastewater treatment, Chap. 1 presents the advantages linked to the application of chemically assisted primary sedimentation (CAPS); this enables energy optimization of wastewater treatment plants and points to the possibility of wastewater as a possible resource. The increase in production of primary sludge obtained in CAPS generates a major production of biogas in anaerobic digestion which can off-set the power used for the treatment process, thus reducing the use of non-renewable energy. In addition, this chapter discusses the use of organic coagulants such as chitosan (which has a much lower cost) in the CAPS process.

In recent years, there has been increasing concern about the release of contaminants such as endocrine disruptors compounds (EDCs), pharmaceuticals and personal care products (PPCPs) into the environment. These contaminants are ubiquitous, persistent and biologically active, and they may cause disruption of endocrine systems as well as affect the hormonal control of development in aquatic and terrestrial biota. Because the majority of emerging compounds are detectable in the environment at concentrations ranging in ng/L to µg/L levels, Chap. 2 discusses the analytical problems related to the analytical detection of pollutants and of their transformation products.

The effluent from urban wastewater treatment plants (UWWTPs) are among the major sources involved in surface water contamination by EDCs/PPCPs. Hundreds of tons of pharmacological substances enter UWWTPs each year, and very likely they would not be degraded by the physical and biological processes, thus contributing to widespread environmental pollution. The likelihood of water contamination with emerging pollutants as a result of discharge of UWWTP effluents depends on several factors. Among them the most important are: (i) the physico-chemical properties of pollutants; (ii) the wastewater treatment technology in use; (iii) the type of activated sludge process, and (iv) the climatic conditions (such as rainfall, temperature and sunlight). Activated sludge has been the most frequently used biological process in wastewater treatment plants. However, its effectiveness in the removal of emerging contaminants has been recently questioned leading to the employment of advanced processes. A number of investigators have addressed this issue over the past decade highlighting the promising role of a special class of oxidation techniques defined as advanced oxidation processes (AOPs). Other studies have focused on membrane bioreactors (MBR) to treat emerging contaminants from water and wastewater. A brief description of MBR principles and technology is presented in Chap. 3, with updated figures showing the distribution of these plants worldwide. Based on literature data, the efficiency of MBR plants for removing trace pollutants is compared with conventional systems. Finally, there is a discussion on the "green" character of MBR technology.

Chapter 4 evaluates the application of Wet Oxidation (WO) for the treatment of aqueous effluents to remove trace pollutants, based on flow rate and organic content in the wastewater effluent.

Chapter 5 is a review of the application of Photo-Fenton process and complementary treatment systems (H_2O_2/UV–C and Fenton's reagent) for the degradation of two industrial pollutant categories with significant endocrine disrupting properties: alkyl phenols (nonyl and octyl phenols) and bisphenol A.

All chapters include fundamentals of the processes investigated as well "green aspects" of technologies that will offer students, technicians, and academics the opportunity to evaluate and select the technologies that lead to better and more sustainable treatment.

Salerno, October 2011   Giusy Lofrano

Acknowledgments

This book was created within the series "Springer Briefs in Green Chemistry for Sustainability" edited by Prof. Sanjay Sharma. To him, my warmest thanks.

My most sincere gratitude goes to all the authors who devoted their precious time to contribute to this volume and to Sonia Ojo, Ilaria Tassistro and their team at Springer Publisher for their valuable support that made this book possible. It has been a pleasure working with all of you.

My thanking to Jeanette Brown, who honored me by signing the Foreword, will never be enough.

I wish to express my most sincere gratitude to Süreyya Meric for all the competence showed over the years and for the scientists and friends who gave me the chance to meet.

I am beholden to Giovanni De Feo whose brilliance, wittiness and vision of the environment informed and inspired me deeply. A special thanks to Ivana Marino who supported me continuously and unconditionally and to Giovanni Pagano who helped me with his encouraging comments.

Last but not least, I am grateful to my family. Your patience and love power my life.

Salerno, October 2011   Giusy Lofrano

Contents

Contributors

Idil Arslan-Alaton Environmental Engineering Department, Istanbul Technical University, Ayazaga Kampusu, Maslak, 34469 Istanbul, Turkey, e-mail: arslanid@itu.edu.tr

Giorgio Bertanza Faculty of Engineering, University of Brescia, via Branze 43, 25123 Brescia, Italy, e-mail: bert@ing.unibs.it

Santiago Esplugas Department of Chemical Engineering, University of Barcelona, Martí i Franquès 1, 08028 Barcelona, Spain, e-mail: santi.esplugas@ub.edu

Giovanni De Feo Department of Industrial Engineering, University of Salerno, Via Ponte don Melillo, 84084 Fisciano, SA, Italy, e-mail: g.defeo@unisa.it

Maurizio Galasso Bierrechimica S.r.l., via Canfora 59/61, 84084 Fisciano, SA, Italy, e-mail: mauriziogalasso@bierrechimica.it

Verónica García-Molina Dow Chem Iberica SL, Dow Water and Process Solutions, Tarragona, Spain, e-mail: vgarciamolina@dow.com

Sabino De Gisi Environmental Engineer, via Contrada Santissimo 35, 83042 Atripalda, AV, Italy, e-mail: sdegisi@unisa.it

Giusy Lofrano Department of Civil Engineering, University of Salerno, via Ponte don Melillo, 84084 Fisciano, SA, Italy, e-mail: glofrano@unisa.it

Anastasia Nikolaou Department of Marine Sciences, University of the Aegean, University Hill, 81100 Mytilene, Lesvos, Greece, e-mail: nnikol@aegean.gr

Tugba Olmez-Hanci Environmental Engineering Department, Istanbul Technical University, Ayazaga Kampusu, Maslak, 34469 Istanbul, Turkey, e-mail: tolmez@itu.edu.tr

Roberta Pedrazzani Faculty of Engineering, University of Brescia, via Branze 43, 25123 Brescia, Italy, e-mail: roberta.pedrazzani@ing.unibs.it

Abbreviations

ADBI	4-Acetyl-1,1-dimethyl-6-tert-butylindane
AHMI	6-Acetyl-1,1,2,3,3,5-hexame-thylindane
AHTN	7-Acetyl-1,1,3,4,4,6-hexamethyl-1,2,3,4-tetrahydronaphthalene
AMBI	5-Amino-6-methyl-2-benzimidazolone
AOPs	Advanced oxidation processes
AOX (RCl)	Adsorbable organically bound halogens
APs	Alkyl phenols
APEOs	Alkyl phenol ethoxylates
ATII	5-Acetyl-1,1,2,6-tetramethyl-3-isopropylindane
BE	Benzoylecgonine
BOD5	Biochemical oxygen demand
BPA	Bisphenol A
BSA	N,O-bis(trimethylsilyl)-acetamide
BSTFA	N,O-bis(trimethylsilyl)-trifluoroacetamide
CAS	Conventional activated sludge
CAPS	Chemically assisted primary sedimentation
CE	Cocaethylene
CEPT	Chemically enhanced primary treatment
CLLE	Continuous liquid–liquid extraction
$C_2O_4^{2-}$	Oxalate ion
$C_2O_4^{\bullet-}$	Oxalate radical
COD	Chemical oxygen demand
DEET	N,N diethyl-m-toluamide
DOC	Dissolved organic carbon
DOM	Dissolved organic matter
DNOM	Dissolved natural organic matter
DPMI	6,7-Dihydro-1,1,2,3,3-pentamethyl-4-(5H)-indanone
ECD	Electron capture detector

EDCs	Endocrine disruptors compounds
EDDP	2-Ethylidine-1,5-dimethyl-3,3-diphenylpyrrolidine perchlorate
EI	Electron impact
EPA	Environmental protection agency
Fe(IV)	Ferrate
FID	Flame ionization detector
FS	Plat-and-frame/flat sheet
GC	Gas chromatography
GCB	Graphitized carbon black
GLKM	General lumped kinetic model
HHCB	1,2,4,6,7,8-Hexahydro-4,6,6,7,8,8-hexamethylcyclopenta-c-2-benzopyrane
HF	Hollow fiber
$HO^{\bullet}$	Hydroxyl radical(s)
$HO_{ss}^{\bullet}$	Steady-state hydroxyl radical concentration
$HO_2^{\bullet}$	Hydroperoxyl radical(s)
HPLC	High performance liquid chromatography
iMBR	immersed MBR configuration
IT	Ion trap
KDIE	Kinetic deuterium isotope effect
L	Organic ligand
LITs	Linear ion traps
LC	Liquid chromatography
LLE	Liquid–liquid extraction
LMCT	Ligand to metal charge transfer
LOEC	Lowest observed effect concentration
LSD	Lysergic acid diethylamide
MBR	Membrabe bio reactors
MCF	Methyl chloroformate
MDA	3,4-Methylenedioxyamphetamine
MDE or MDEA	Methylenedioxy-ethylamphetamine
MDMA	3,4-Methy-138 lenedioxymetamphetamine hydrochloride
MIPs	Molecularly imprinted polymers
MLVSS	Mixed liquor volatile suspended solids
MS/MS	Tandem mass spectrometric
MSTFA	N-Methyl-N trimethylsilyltrifluoroacetamide
MT	(Multi)tubular
MTBSTFA	N-(Tert-butyldimethylsilyl)-N Methyltrifluoroacetamide

NCI	Negative chimica ionization
NOEC	No observed effect concentration
NP	4-Nonylphenol
NPEOs	Nonylphenol ethoxylates
NPEO-9	Nonylphenol ethoxylate-9
NPEO-10	Ten-fold ethoxylated nonylphenol
NP1EO	4-Nonylphenol monoethoxylate
NP2EO	4-Nonylphenol diethoxylate
$O_2^{\bullet-}$	Superoxide radical(s)
OECD	Organization for economic cooperation and development
OP	Octylphenol
OPEOs	Octylphenol ethoxylates
PCPs	Personal care products
PPCPs	Pharmaceuticals and personal care products
PFA	Pentafluoropropionic acid anhydride
QqQ	Quadrupole
QqLIT	Quadrupole–linear ion trap
QqTOF	Quadrupole–time of flight
RLU	Relative light units
sMBR	sidestream MBR configuration
SPE	Solid phase extraction
SPME	Solid phase micro extraction
SRT	Solid retention time
TBS	Tert-butyldimethylsilyl
THC	D9-tetrahydrocannabinol
TKN	Total Kjeldahl Nitrogen
TMS	Trimethylsilyl
TMS-DEA	N,N-diethyltrimethylamine
TOC	Total organic carbon
TOD	Total oxygen demand
TrBA	Tri-n-butylamine
TMS-DEA	N,N-diethyltrimethylamine
TSS	Total sospende solids
UV	Ultra violet
UWWTP	Urban wastewater treatment plants
VSS	Volatile suspended solids
WAO	Wet air oxidation
WO	Wet oxidation

WPO	Wet peroxidation
WWTPS	Wastewater treatment plants
YES	Yeast estrogen screen
4-CP	Chlorophenol
4-OP	4-tert-octylphenol
4-NP	4-nonylphenol

Chapter 1
Chemically Assisted Primary Sedimentation: A Green Chemistry Option

Giovanni De Feo, Sabino De Gisi and Maurizio Galasso

Abstract Chemically Assisted Primary Sedimentation (CAPS) consists of adding chemicals in order to increase the coagulation, flocculation and sedimentation of raw urban wastewater. The CAPS process can be developed in order to increase the efficacy of primary sedimentation as well as avoid any interference with biological treatment processes. The application of CAPS is particularly suitable as a technique for the upgrading of urban wastewater treatment plants (UWWTPs). In fact, CAPS does not require any further significant structural intervention (so saving investment costs and territory portions). The aim of this contribution is to emphasise the role of CAPS as a green chemistry option available in a UWWTP. In particular, a specific aim of the chapter is to focus the attention on the energetic importance of CAPS due to its capacity to increase the production of the primary sludge and consequently the energy production with an anaerobic digestion treating separately primary and secondary sludge. The energetic convenience and "green" propensity of the application of CAPS is discussed by means of the presentation of a paradigmatic case study containing economic evaluations, as well.

G. De Feo (✉)
Department of Industrial Engineering, University of Salerno,
Via Ponte don Melillo 1, 84084 Fisciano (SA), Italy
e-mail: g.defeo@unisa.it

S. De Gisi
Via Contrada Santissimo 35, 83042 Atripalda (AV), Italy
e-mail: sdegisi@unisa.it

M. Galasso
Bierrechimica S.r.l., Via Canfora, 59/61,
84084 Fisciano (SA), Italy
e-mail: mauriziogalasso@bierrechimica.it

G. Lofrano (ed.), *Green Technologies for Wastewater Treatment*,
SpringerBriefs in Green Chemistry for Sustainability,
DOI: 10.1007/978-94-007-1430-4_1, © De Feo, De Gisi, Galasso 2012

Keywords CAPS · CEPT · Economical analysis · Sludge management system · Urban wastewater

1.1 Introduction

Chemically Assisted Primary Sedimentation (CAPS) or Chemically Enhanced Primary Treatment (CEPT) consists of adding chemicals in order to increase the coagulation, flocculation and sedimentation of raw urban wastewater. The CAPS process can be developed in order to improve the efficacy of primary sedimentation as well as avoid any interference with biological treatment processes. The removal percentages, due to the CAPS application on urban wastewater, are between 60 and 90% of the total suspended solids (TSS), 40–80% of the 5-day biochemical oxygen demand (BOD_5), 30–70% of the chemical oxygen demand (COD), 65–95% of the phosphorus, and 80–90% of bacteria [1, 2]. On the other hand, simple primary sedimentation may remove between 50–70% of TSS, 25–40% of BOD_5, 5–10% of phosphorus, and 50–60% of pathogens [1, 2].

CAPS is not an innovative process but it allows the energy optimisation of a large wastewater treatment plant as well as it makes possible to consider the wastewater as a possible resource (see Table 1.1).

The increase in production of primary sludge obtained in the CAPS process generates a major production of biogas in anaerobic digestion and therefore a lower energy demand. In addition, the use of organic coagulants (such as chitosan) at affordable cost is the most interesting aspect for the implementation of CAPS. The primary sludge produced has a higher content of organic substances (in terms of dry) and, thus, a potential energy greater than in the case of the primary natural sedimentation. In addition, the nonuse of metal coagulants (such as ferric chloride, aluminium sulphate, etc.) prevents damages to the biological processes of the anaerobic digestion.

The application of CAPS is particularly suitable as a technique for the upgrading of urban wastewater treatment plants (UWWTPs). In fact, CAPS does not require any further significant structural intervention. The optimal requirements for its application in a UWWTP are the presence of both a primary sedimentation tank and a sludge anaerobic digestion basin. In general, the application of CAPS may be useful in the following situations:

- in a UWWTP where the activated sludge process is designed with a high load (Food/Microorganisms ratio)
- in a UWWTP subject to a strong seasonal variation of the inlet hydraulic and organic load, typical of coastal regions [3, 4]
- in a UWWTP potentially affected by management problems in presence of industrial wastewater (e.g., synthetic and natural tannins in the studied WWTP).

Table 1.1 Current aspects of chemically assisted primary sedimentation process

Main aspects	Sub-objectives
Energy efficiency in urban wastewater treatment plant	Changing of WWTP flow chart with the introduction of innovative and/or consolidated processes;
	Improving energy performance of electro-mechanical equipment (i.e., pumps)
Wastewater as a possible resource	Increasing biogas production in the anaerobic digestion of primary sludge
	Possible reuse of biological sludge (secondary sludge) in agriculture
	Possible reuse of treated wastewater

The aim of this contribution is to emphasise the role of the chemically assisted primary sedimentation as a green chemistry option available in a UWWTP. In particular, a specific aim of the chapter is to focus the attention on the energetic importance of CAPS due to its capacity to increase the production of the primary sludge and consequently the energy production with an anaerobic digestion treating separately primary and secondary sludge. First of all, the manuscript presents the CAPS as a green technology as well as discusses the major advantages deriving from its application in a UWWTP. Secondarily, the energetic convenience of the application of CAPS is discussed by means of the presentation of a paradigmatic case study containing economic evaluations, as well.

1.2 CAPS as a Green Technology

The advantages of the application of the CAPS process are in the production of primary sludge with a subsequent increase in the production of biological gas from the anaerobic digestion phase [5, 6] as well as the reduction of the food/microorganisms ratio (F/M) with the subsequent reducing of the energy costs for the activated sludge process. The improvement of both the chemical characteristics of the secondary (biological) sludge (with its possible reuse in agriculture) and final effluent quality (due to a smaller production of non-settleable solids) are further advantages [7, 8]. Disadvantages are related to the pH alteration with possible damage to the activity of the microorganisms in the biological stage, the chemicals costs as well as a slight complication of the process scheme [2, 9, 10]. Moreover, the greater COD reduction in the primary sedimentation may have negative effects on nitrogen removal, but this aspect is site-specific. In fact, it depends by the value of the compliance limit as well as the COD/TKN ratio in the wastewater (TKN, Total Kjeldahl Nitrogen).

The success of the CAPS application in UWWTPs depends on the particles size distribution of COD and their related biodegradability.

Table 1.2 Examples of effluents treated by means of a coagulation/flocculation process using chitosan—Reprinted from Ref. [18] with kind permission of Elsevier

Effluent	Reference(s)[a]
Food, seafood and fish processing wastes	Bough (1975); Bough (1976); No and Meyers (1989); Guerrero et al. (1998); Savant and Torres (2000); Fernandez and Fox (1997)
Wastewater from milk processing plant	Chi and Cheng (2006)
Brewery wastewater	Cheng et al. (2005)
Surimi wash water	Wibowo et al. (2007); Savant and Torres (2003)
Inorganic suspensions (bentonite, kaolinite)	Roussy et al. (2005); Roussy et al. (2005); Roussy et al. (2004); Huang and Chen (1996); Divakaran and Pillai (2004)
Bacterial suspensions	Xie et al. (2002); Strand et al. (2001); Strand et al. (2003)
Effluents containing humic substances	Bratskaya et al. (2002); Bratskaya et al. (2004)
Effluents containing dyes	Guibal et al. (2006); Guibal and Roussy (2007)
Pulp and paper mill wastewater	Rodrigues et al. (2008); Wang et al. (2007); Chen et al. (2006), Sanghi and Bhattacharya (2005); Ganjidoust et al. (1997)
Olive oil wastewater	Ahmad et al. (2006); Ahmad et al. (2004); Meyssami and Kasaeian (2005); Rizzo et al. (2008)
Oil-in-water emulsions	Bratskaya et al. (2006)
Aquaculture wastewater	Chung (2006)
Effluent containing metal ions	Gamage and Shahidi (2007); Assaad et al. (2007); Wu et al. (2008)
Effluent containing phenol derivatives	Wada et al. (1995)
Partially purified sewage	Zeng et al. (2008)
Brackish water	Divakaran and Pillai (2002)
Raw drinking water	Rizzo et al. (2008)

[a] All references cited are reported in Ref. [18]

In relation to the particle size distribution of COD, Marani et al. [5] and Ramadori et al. [6] carried out a study on the Roma-Nord sewage treatment plant in Italy (780,000 equivalent inhabitants; 354,000 m^3/d), showing that the COD in sewage is predominantly associated with settleable and supra-colloidal particles, with each size range including about 40% of total COD. These results are comparable to other findings present in literature and related to European cases [11–16]. According to Odegaard's studies [9] about 25% of COD was related to particles with a size <0.08 μm, 15% of COD was found to be appearing as colloidal (0.08–1.0 μm), about 25% as supra-colloidal (1.0–100 μm) and about 35% as settleable (>100 μm) particles.

In relation to the biodegradability of different size fractions, a large fraction of COD associated with supra-colloidal particles is characterised by slow degradability [5, 6]. This result suggests that, the elimination of slowly degradable particles prior to the aerobic biological treatment promotes a more effective utilisation of the biological treatment capacity [5, 6]. Moreover, Mininni et al. [8] highlighted the convenience of separating primary sludge from the secondary one before treatment and disposal. In fact, as it is well known, primary and secondary sludges have different physical properties: primary sludge can be thickened, digested and mechanically dewatered much better than its mixture with the secondary sludge [17].

Moreover secondary sludge, if treated separately from the primary sludge, may be used directly in agriculture (considering its relatively high content of nitrogen and phosphorus as well as negligible presence of pathogens and micropollutants). Primary sludge may be incinerated after the typical operating sequence of gravity thickening, anaerobic digestion and mechanical dewatering. In the light of the above considerations, the application of CAPS to a sewage treatment plant with a capacity between 100,000 and 500,000 equivalent inhabitants can be based on a sludge management system with the separation of the treatment and disposal of the primary and secondary sludge.

The use of organic coagulants in place of conventional coagulants, such as metal salts, is another interesting aspect of CAPS. Since the primary sludge produced with CAPS has a dry content of organic matter higher than that of inorganic coagulants, a greater production of biogas can be obtained. Numerous examples of using organic coagulants in the fields of wastewater treatment and drinking water are present in the literature. As matter of fact, Table 1.2 shows the fields of application of chitosan, one of the most used organic coagulants, according to the most recent literature. Chitosan is a partially deacetylated polymer obtained from the alkaline deacetylation of chitin, a biopolymer extracted from shellfish sources. It has received a great deal of attention in the last decades in water treatment processes for the removal of particulate and dissolved contaminants. However, at present, the application of chitosan and other organic coagulants in commerce is limited by the high cost.

In the light of both the sustainable development paradigm and energetic shortage world crisis, a paradigmatic case study is shown in the next paragraph in order to emphasise the environmental convenience in using a chemically assisted primary sedimentation process in a urban wastewater treatment plant as a green chemistry option.

1.3 A Paradigmatic Case Study

1.3.1 Case Study Materials and Methods

1.3.1.1 The Wastewater Treatment Plant Under Study

The wastewater treatment plant under study is situated in the city of Avellino, in Southern Italy, in the Campania region. It had an organic load corresponding to around 140,000 equivalent inhabitants with a specific production of BOD_5 equal to

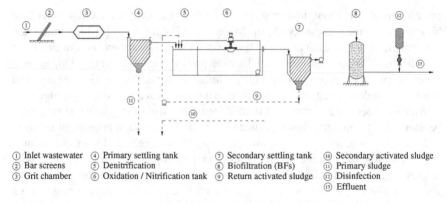

① Inlet wastewater   ④ Primary settling tank   ⑦ Secondary settling tank   ⑩ Secondary activated sludge
② Bar screens   ⑤ Denitrification   ⑧ Biofiltration (BFs)   ⑪ Primary sludge
③ Grit chamber   ⑥ Oxidation / Nitrification tank ⑨ Return activated sludge   ⑫ Disinfection
                                                                                  ⑬ Effluent

Fig. 1.1 Water processing flow diagram considered in this study

60 g_{BOD5}/inhabitant/d. The scheme of the plant included the following units for the treatment of the inlet urban wastewater (Fig. 1.1): bar screens, grit chamber, primary settling tank, denitrification, oxidation/nitrification tank, secondary settling tank, submerged biological filter (BFs) and disinfection.

While, the sludge treatment scheme included the following units: gravity thickener tank, anaerobic sludge digestion tank, chemical conditioning and sludge dewatering with centrifuge and belt-filter press. The sludge management system of the considered WWTP is based on a mixture of primary and secondary sludge entering the sludge line. While, as stated above, two hypothesis were considered in this paper (Fig. 1.2): combined treatment (anaerobic digestion) and traditional disposal of the primary and secondary sludge (Fig. 1.2a); anaerobic digestion of the primary sludge and direct reuse of the secondary sludge in agriculture (Fig. 1.2b). The knowledge of the two flow charts is important in order to correctly understand the results of the mass balance successively presented and discussed.

1.3.1.2 Analytical Methods and Inlet Wastewater Characteristics

The chemical–physical characterisation of the inlet raw wastewater was carried out according to *Standard Methods* [19]. A series of laboratory analysis were conducted in the period from May to December 2005 in relation to the largest hydraulic and organic daily loads (Table 1.3).

1.3.1.3 Phases of the Experimentation

The experimental activity was developed according to the following phases:

- analysis of the literature dealt with CAPS as well as data regarding the WWTP under study

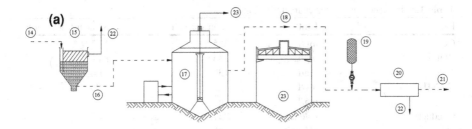

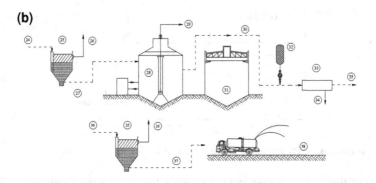

(14) Primary and secondary sludge
(15) Gravity thickener tank
(16) Sludge
(17) Anaerobic sludge digestion tank
(18) Stabilized sludge

(19) Chemical conditioning
(20) Dewatering with centrifuge and belt-filter press
(21) Dewatered biosolids flow to disposal
(22) Underflow to plant influent
(23) Biogas to gasometer

(24) Primary sludge
(25) Gravity thickener tank
(26) Underflow to plant influent
(27) Primary thickened sludge
(28) Anaerobic sludge digestion tank
(29) Biogas to gasometer
(30) Stabilized primary sludge
(31) Gasometer

(32) Chemical conditioning
(33) Dewatering with centrifuge and belt-filter press
(34) Underflow to plant influent
(35) Dewatered biosolids flow to disposal
(36) Secondary sludge
(37) Secondary thickened sludge
(38) Sludge for agricultural utilisation

Fig. 1.2 Solid processing flow diagram considered in this study: **a** combined treatment (anaerobic digestion) and traditional disposal of the primary and secondary sludge; **b** anaerobic digestion of the primary sludge and direct reuse of the secondary sludge in agriculture

- jar tests on the inlet raw wastewater in order to evaluate the behaviour of eight coagulants (aluminium sulphate, Ecofloc 614, Ecofloc 616, Ecofloc CP, Ecofloc SA7, ferric chloride, polyaluminum chloride (PACl), and sodium aluminate) for five fixed doses (10, 20, 30, 40 and 50 mg/L) with 0.1 mg/L of an anionic polyelectrolyte

Table 1.3 Influent wastewater characterisation

Parameter	Unit	Average value	Standard deviation
COD	mg/L	365.4	127.3
BOD$_5$	mg/L	243.6	84.8
TSS	mg/L	130.7	61.4
N–NH$_4^+$	mg/L	32.7	7.9
Cl$^-$	mg/L	76.3	13.8
Conductivity	μs/cm	757.4	84.3
pH	pH unit	8.0	0.1
Alkalinity	°F	30.1	2.7
Max feed flowrate	m^3/d	27,205	4,053.2

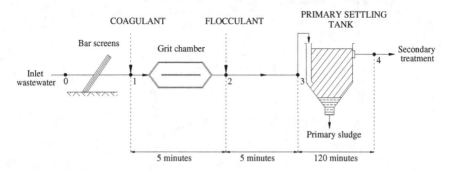

Fig. 1.3 Flow chart of CAPS upgrading in the studied WWTP

- selection of the best coagulant and the relative dosage with the application of a "practical multi-criteria procedure"
- chemical–physical characterisation of natural sludge and CAPS sludge, using the coagulant chosen in the previously phase
- data processing
- verification of the economical applicability of CAPS (with 40 mg/L of Ecofloc CP and 0.1 mg/L of anionic polyelectrolyte) to the WWTP under study and comparison of the results obtained with those without the application of CAPS.

1.3.1.4 Choice of the Best Coagulant and its Relative Dosage

In order to upgrade the conventional primary treatment unit to a CAPS facility, a chemical coagulant must be added as well as an optional flocculent, as shown in Fig. 1.3. A series of jar tests had to be carried out in order to determine which coagulant was the most suitable for the CAPS of raw wastewater [1].

Eight coagulants were evaluated: aluminium sulphate, Ecofloc 614, Ecofloc 616, Ecofloc CP, Ecofloc SA7, ferric chloride, polyaluminum chloride (PACl), and sodium aluminate. The coagulants were evaluated at doses of 10, 20, 30, 40 and 50 mg/L. The raw wastewater was dosed with a coagulant, rapidly mixed for

5 min at 120 rpm, followed by a slow mixing of 5 min at 30 rpm after a fixed 0.1 mg/L dose of an anionic polyelectrolyte had been added. The obtained results showed that the best coagulant was Ecofloc CP at 40 mg/L [1].

The principal advantages obtained with this coagulant were the following: maximisation of COD percentage removal (41.1%) in the primary sedimentation (minimisation of the organic load influent to the secondary biological treatment); minimisation of adverse effects on the biological processes (i.e., microorganisms inhibition) of the secondary treatment by controlling the pH percentage variation between primary sedimentation and CAPS [1].

1.3.1.5 Evaluation of the Sludge Production and Other CAPS Aspects

In order to simulate the application of CAPS to the studied WWTP, a period of 8 months from May to December 2005 was set. In particular, a period of 6 h for every day (corresponding both to the maximum hydraulic and organic loading) was considered. All the technical aspects estimated in this test were referred to the set period. A correct comparison between the set-up with the natural sedimentation (in the following named "Natural Set-up") and the two set-ups with the CAPS application (in the following named "Assisted Set-up 1" and "Assisted Set-up 2") was carried out. The "Assisted Set-up 1" was based on a combined treatment (anaerobic digestion) and a traditional disposal of the primary and secondary sludge. The "Assisted Set-up 2" instead was based on an anaerobic digestion of the primary sludge and a direct reuse of the secondary sludge in agriculture.

The following technical aspects were taken into account: production of primary sludge, production of secondary biological sludge, production of biological gas from anaerobic digestion of sludge, production of dewatered sludge and finally, consumption of energy related to both aeration and nitrification phases.

The value of the primary sludge production, both for the "Natural Set-up" and the "Assisted Set-ups", was determined by Eq. (1.1):

$$q_{ps} = S_{set} \cdot q \tag{1.1}$$

where

- q_{ps} = volumetric flowrate of the primary sludge production, cm^3_{sludge}/d;
- S_{set} = value of settleable solids (suspended solids that will settle out of suspension within a period of 2 hours), $cm^3_{sludge}/m^3_{wastewater}$;
- q = maximum value of the wastewater flowrate inlet to the WWTP, $m^3_{wastewater}/d$.

By means of a series of jar tests with Ecofloc CP (40 mg/L) and 0.1 mg/L of anionic polyelectrolyte, it was possible to estimate the relationship between Total Suspend Solids (TSS) and S_{set}. Therefore, the values of S_{set} were determined from the corresponding values of TSS both for the "Natural Set-up" and "Assisted Set-ups". The production of secondary activated sludge was evaluated by Eq. (1.2) [20]:

Table 1.4 Values assumed for the evaluation of dewatered sludge for the "Natural Set-up" (natural primary sedimentation)

Parameter	Total dry solids (TS)	Volatile solids as percentage of TS
Primary sludge (%)	3	75
Secondary activated sludge (%)	1	75
Primary and secondary sludge (%)	1.7	75
After gravity thickener (%)	4	75
After anaerobic digestion (%)	4	–
After dewatering (%)	25	–

Table 1.5 Values assumed for the evaluation of dewatered sludge for the "Assisted Set-up 1" (CAPS, combined anaerobic digestion and traditional disposal of the primary and secondary sludge)

Parameter	Total dry solids (TS)	Volatile solids as percentage of TS
Primary sludge (%)	3	70
Secondary activated sludge (%)	1	75
Primary and secondary sludge (%)	2	72.4
After gravity thickener (%)	4	72.4
After anaerobic digestion (%)	4	–
After dewatering (%)	25	–

$$I = 10^{(0.072 + 0.229 \cdot \mathrm{Log}\, F/M)} \tag{1.2}$$

where

- I = sludge production index, $kg_{produced}/kg_{removed\ BOD5}$
- F/M = food to microorganisms ratio, $kg_{BOD5}/kg_{MLVSS}/d$.

The value of the F/M ratio was calculated for the three aeration tanks present in the studied WWTP called "Putignano" (with a volume of 3,720 m^3), "Exim" (with a volume of 1,860 m^3) and Degrémont-Ondeo (with a volume of 3,000 m^3), respectively. Every day, the production of secondary biological sludge was evaluated for each aeration tank considered. This evaluation was performed on the basis of the values of the concentration of the mixed liquor volatile suspended solids (MLVSS) and in relation to the average COD percentage removal in the primary sedimentation both for the "Natural Set-up" and "Assisted Set-ups", equal to 23% (average between 20 and 25%) and 41.1% (value estimated in the laboratory with jar tests), respectively.

The consumption of energy both for the oxidation and nitrification phases was assumed to be equal to 2 kWatt × h for each kg of inlet BOD$_5$. A mass balance on the WWTP sludge line was performed in order to calculate the production of dewatered sludge for the three considered set-ups. As a matter of fact, Tables 1.4, 1.5 and 1.6 show the values assumed by the principal parameters for the "Natural Set-up", "Assisted Set-up 1" and "Assisted Set-up 2, respectively. In relation to the anaerobic digestion of sludge, a percentage removal of volatile suspended solids (VSS) of 65%

Table 1.6 Values assumed for the evaluation of dewatered sludge for the "Assisted Set-up 2" (CAPS, anaerobic digestion of the primary sludge and direct reuse of the secondary sludge in agriculture)

Parameter	Total dry solids (TS)	Volatile solids as percentage of TS
Primary sludge (%)	3	70
Secondary activated sludge (%)	1	75
After gravity thickener (%)	5	70
After anaerobic digestion (%)	5	–
After dewatering (%)	25	–

Table 1.7 Values of unit costs and benefit considered in this study

Cost	Unit	Value
Energy for the aeration and nitrification phases	€/kWatt/h	0.15
Dewatering biosolids flow to disposal	€/t	100.00
Coagulant–Ecofloc CP	€/kg	0.25
Flocculent	€/kg	2.50
Benefit	Unit	Value
Biogas production	€/Nm^3	0.33

Nm^3 = normal cubic meter of biogas at pressure of 760 mmHg and temperature of 20 °C
€ = Euro

was assumed and a biological gas production of 1,000 L for every removed kilogram of volatile suspended solids, for the two considered set-ups [20, 21]. Finally, Table 1.7 shows the values assumed for the economic variables.

1.3.2 Case Study Results and Discussion

1.3.2.1 Evaluation of the Relationship Between TSS and Settleable Solids and Chemical–Physical Characterisation of the Sludge Produced

In this paragraph the results of a series of jar tests carried out with 40 mg/L of Ecofloc CP and 0.1 mg/L of anionic polyelectrolyte are presented. As shown in Fig. 1.4, the relationship TSS-S_{set} was determined both for the "Natural Set-up" and "Assisted Set-ups". While, Table 1.8 proposes the chemical characterisation of the sludge for the considered set-ups (the concentrations are expressed in mg/kg of dry solids). An increase of iron, aluminium and manganese concentrations was observed in the sludge produced by the CAPS application (with 40 mg/L of Ecofloc CP and 0.1 mg/L of anionic polyelectrolyte) due to the chemical nature of Ecofloc CP. In particular, the iron and aluminium increases were about 144% and 30%, respectively. It is well known that the biogas production from the anaerobic phase may be inhibited by high concentrations of iron and aluminium present in the inlet sludge [21, 22]. Moreover, it is worth pointing out that the CAPS application

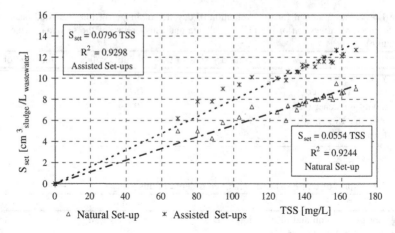

Fig. 1.4 Relationships TSS-S$_{Set}$ for the "Natural Set-up" and "Assisted Set-ups"

regarded only a period of 6 h for every day (corresponding both to the maximum hydraulic and organic loading). In this case, the metal concentration values corresponding to the "Assisted Set-ups" could be smaller than the respective values in Table 1.8. Finally, as shown in Table 1.8, the metal concentrations of the "natural sludge" were comparable with the data present in literature and relating to the characteristics of sludge from Italian urban WWTP [23].

1.3.2.2 Evaluation of the Primary and Secondary Biological Sludge

Referring to the studied time period, the primary and secondary sludge productions were evaluated and the obtained results are shown in Fig. 1.5. It is well known that the use of chemical–physical treatment allows an increase of primary sludge production [10, 21]. As a matter of fact, an increase from 8,710.5 m^3 to 12,515.4 m^3 (+43.3%) was registered in the study. While, the reduction of the F/M ratio allowed a smaller secondary sludge production in the "Assisted Set-ups" compared to the "Natural Set-up" with a decrease of 30.1%. Finally, the total sludge production (primary and secondary sludge) in the "Assisted Set-ups" was slightly less than the corresponding production for the "Natural Set-up", as shown in Fig. 1.5.

1.3.2.3 Biological Gas Production and Consumption of Energy for the Aeration and Nitrification Phases

The consumption of energy for the final biological treatment as well as the results of the biological gas production from the anaerobic sludge digestion for the three studied set-ups are shown in Fig. 1.6. Relating to the "Assisted Set-ups", a decrease of the electricity consumption of about 24% was observed, due to the

Table 1.8 Chemical–physical characterisation of primary sludge for the "Natural Set-up" and the "Assisted Set-ups" (40 mg/L of Ecofloc CP, 0.1 mg/L of flocculent)

Metals	Sludge		
	Natural Set-up	Assisted Set-up	Variation (%)
Lead (mg/kg$_{SS}$)	2.00	1.72	−13.0
Cadmium (mg/kg$_{SS}$)	<1.00	<1.00	–
Nickel (mg/kg$_{SS}$)	14.88	10.10	−32.0
Chrome (mg/kg$_{SS}$)	44.10	34.48	−21.0
Copper (mg/kg$_{SS}$)	117.97	73.89	−37.0
Aluminium (mg/kg$_{SS}$)	3,005.44	3,899.01	+29.0
Iron (mg/kg$_{SS}$)	720.51	1,757.39	+143.0
Antimony (mg/kg$_{SS}$)	0.00	0.00	–
Vanadium (mg/kg$_{SS}$)	3.09	2.34	−24.0
Manganese (mg/kg$_{SS}$)	58.08	72.66	+25.0

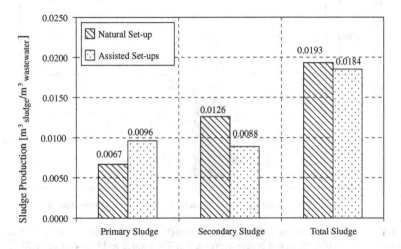

Fig. 1.5 Sludge production for the "Natural Set-up" and "Assisted Set-ups"

smaller F/M ratio of the biological treatment in comparison with the "Natural Set-up" which is produced by the increase of COD percentage removal in the primary sedimentation with the CAPS application [1, 21]. Moreover, passing from the "Natural Set-up" to the "Assisted Set-up 1" an increase of 10.9% in the biological gas production was registered. While passing from the "Natural Set-up" to the "Assisted Set-up 2" a decrease of 17.9% in the same gas production was estimated. This result, for the "Assisted Set-up 2" was due to a minor sludge flow entering the anaerobic digester, although the primary sludge is more suitable than the mixture of primary and secondary sludge for biogas production [17].

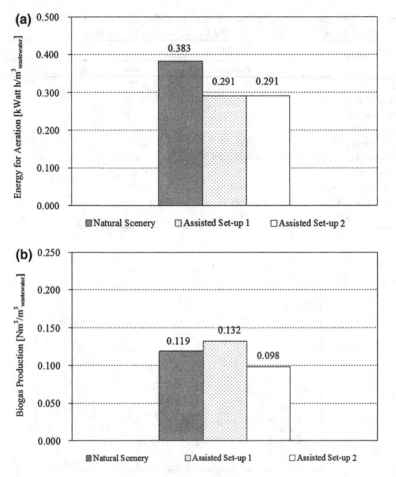

Fig. 1.6 Energy for aeration (**a**) and biogas production (**b**) for the "Natural Set-up" and "Assisted Set-ups" (Nm^3 = cubic metres of biogas at pressure of 760 mmHg and temperature of 20 °C)

1.3.2.4 Evaluation of the Economical Applicability of the CAPS

The results of the performed economic analysis are summarised in Table 1.9 in terms of costs to sustain and benefits obtained. The costs included the following areas: cost of dewatered biosolids to disposal, cost for energy aeration and cost of coagulant and flocculent. On the other hand, only biogas production was considered as a benefit. It is important to point out that in the performed evaluation other running costs usually considered for a WWTP such as personnel costs, maintenance and administration, as suggested by Austermann-Haun et al. [24], were not taken into account because they are not necessary for a comparison among the set-ups. The results of the economic analysis were expressed in €/m^3 coherently with

Table 1.9 Cost-Benefit analysis for the "Natural Set-up" and "Assisted Set-ups" considered

Value (€/m³)	Set-up		
	Natural	Assisted 1[a]	Assisted 2[b]
Dewatered biosolids to dispose	0.064	0.070	0.055
Energy for aeration	0.057	0.044	0.044
Coagulant	0	~0	~0
Flocculent	0	0.003	0.003
Biogas production	0.039	0.044	0.032
Total cost [c]	0.121	0.116	0.101
Total benefit	0.039	0.044	0.032
Cost–Benefit	0.082	0.072	0.069

[a] Assisted set-up based on a mixture of primary and secondary sludge entering the sludge line of the studied WWTP

[b] Assisted set-up based on the separation of the treatment and disposal of primary and secondary sludge

[c] The total cost included the following topics: cost of dewatered biosolids to dispose, cost for energy aeration, cost of coagulant and flocculent

€ = Euro

other authors [25–27]. Relating to the dewatered biosolids to dispose, a decrease of about 14% between the "Natural Set-up" and the "Assisted Set-up 2" was obtained due to the hypothesis of the use of secondary biological sludge in agriculture. Moreover, as shown in Table 1.9, the obtained value for the "Assisted Set-up 2" was less than the same for the "Assisted Set-up 1".

In relation to the energy for aeration, a decrease of about 22.8% between the "Natural Set-up" and the "Assisted Set-up 2" was obtained. Moreover, this value was the same for the two assisted set-ups, as shown in Table 1.9 (the two assisted set-ups were both based on a percentage COD reduction of about 41% in the primary treatment). The cost for the coagulant was about 519 € in the period of study considered. This value was next to zero in relation to the volume of the inlet wastewater (Table 1.9). While, the flocculent cost was about 0.003 €/m³ for the "Assisted Set-ups".

In relation to the theoretical biogas production, a decrease of about 17.9% for the "Assisted Set-up 2" was obtained in comparison to the "Natural Set-up". This result depends by the minor sludge flowrate entering the anaerobic digestion phase although only the primary sludge is better than the mixture of primary and secondary sludge. Moreover, a higher benefit due to the production of biogas was found for the "Assisted Set-up 1".

Finally, as shown in Table 1.9, the following values were obtained in terms of the difference between costs and benefits: 0.082 €/m³ for the "Natural Set-up", 0.072 €/m³ for the "Assisted Set-up 1" and 0.069 €/m³ for the "Assisted Set-up 2". Therefore, passing from the "Natural Set-up" to the "Assisted Set-up 1" a percentage saving of 12.2% was obtained. While passing from the "Natural Set-up" to the "Assisted Set-up 2" a percentage saving of 15.9% was estimated. These values justify the economic applicability of the CAPS at the studied wastewater treatment plant, particularly emphasising economically the choice of a sludge management system based on a separation of primary and secondary sludge treatment and disposal.

1.3.2.5 Case Study Conclusions

The following particular outcomes on the CAPS process based on the results obtained can be stated:

- the reduction of the F/M ratio on the final biological treatment allowed a smaller secondary sludge production in the "Assisted Set-ups" (chemically assisted primary sedimentation) compared to the "Natural Set-up" (natural primary sedimentation) with a decrease of 30.1% and a consequent decrease of the electricity consumption of about 24%
- passing from the "Natural Set-up" to the "Assisted Set-up 1" (combined anaerobic digestion and traditional disposal of the primary and secondary sludge) an increase of 10.9% in the biological gas production was registered
- passing from the "Natural Set-up" to the "Assisted Set-up 2" (anaerobic digestion of the primary sludge and reuse of the secondary sludge in agriculture) a decrease of 17.6% in the same gas production was estimated
- passing from the "Natural Set-up" to the "Assisted Set-up 1" an economic saving of 12.2% was obtained
- passing from the "Natural Set-up" to the "Assisted Set-up 2" an economic saving of 15.9% was estimated.

1.4 Concluding Remarks

The following general outcomes on the CAPS as a green chemistry option can be stated:

- the economic analysis developed with a paradigmatic case study emphasised the applicability of the CAPS process for the treatment of urban wastewater, particularly emphasising the choice of a sludge management system based on a separation of primary and secondary sludge treatment and disposal
- since CAPS does not require further structural interventions, it is particularly suitable as a technique for the upgrading of urban WWTP, avoiding the construction of new units or even new plants (so saving investment costs and territory portions)
- application of CAPS may be useful in numerous situations (in a WWTP where the activated sludge process is designed with a high load; in a WWTP subject to a strong seasonal variation; in a WWTP for the treatment of urban wastewater that could be affected by management problems due to the presence of industrial wastewater)

- principal CAPS "green" effects relate to the increase in the production of primary sludge, lowering of the food/microorganisms ratio (F/M), reduction of the production of secondary biological sludge, reduction of the energy costs due to a lower F/M and, finally, increase in the production of biological gas from the anaerobic digestion phase.

References

1. De Feo G, De Gisi S, Galasso M (2008) Definition of a practical multi-criteria procedure for selecting the best coagulant in a chemically assisted primary sedimentation process for the treatment of urban wastewater. Desalination 230:229–238
2. Poon CS, Chu CW (1999) The use of ferric chloride and anionic polymer in the chemically assisted primary sedimentation process. Chemosphere 39(10):1573–1582
3. Odegaard H, Balmer P, Hanaus I (1987) Chemical precipitation in highly loaded stabilization ponds in cold climates. Water Sci Technol 19(12):71–77
4. Odegaard H (1989) Appropriate technology for wastewater treatment in coastal tourist area. Water Sci Technol 21(1):1–17
5. Marani D, Renzi V, Ramadori R, Braguglia CM (2004) Size fractionation of COD in urban wastewater from a combined sewer system. Water Sci Technol 50(12):79–86
6. Ramadori R, Marani D, Renzi V, Passino R, Di Pinto AC (2005) Rethinking sewage treatment by enhancing primary settling with low-dosage lime. Water Sci Technol 52(10/11):185–192
7. Chao AC, Keinath TM (1979) Influence of process loading intensity on sludge clarification and thickening characteristics. Water Res 13:1213–1223
8. Mininni G, Braguglia CM, Ramadori R, Tomei MC (2004) An innovative sludge management system based on separation of primary and secondary sludge treatment. Water Sci Technol 50(9):145–153
9. Odegaard H (1998) Optimised particle separation in the primary step of wastewater treatment. Water Sci Technol 37(10):43–53
10. Vesilind PA (2003) Wastewater treatment plant design. Water Environment Federation, Alexandria
11. Balmat JL (1957) Biochemical oxidation of various particulate fractions of sewage. Sew Ind Wastes 29(7):757
12. Heukelekian H, Balmat JL (1959) Chemical composition of the particulate fractions of domestic sewage. Sew Ind Wastes 31(4):413
13. Munch R, Hwang CP, Lackie TH (1980) Wastewater fractions add to total treatment picture. Water Sew Works 127:49–54
14. Neis U, Tiehm A (1997) Particle size analysis in primary and secondary waste water effluents. Water Sci Technol 36(4):151–158
15. Odegaard H (1992) Norwegian experiences with chemical treatment of raw wastewater. Water Sci Technol 25(12):255–264
16. Richert DA, Hunter JV (1971) General nature of soluble and particulate organics in sewage and secondary effluents. Water Res 5(7):421
17. Kopp J, Dichtl N (2001) Influence of the free water content on the dewaterability of sewege sludges. Water Sci Technol 44(10):177–183
18. Renault F, Sancey P, Badot M, Crini G (2009) Chitosan for coagulation/flocculation processes—an eco-friendly approach. Eur Polym J 45:1337–1348
19. APHA (1995) Standard methods for the examination of water and wastewater, 19th edn. American Public Health Association/American Water Works Association/Water Environment Federation, Washington

20. Masotti L (1993) Depurazione delle acque, tecniche ed impianti per il trattamento delle acque di rifiuto, 2nd edn. Edizioni Calderini, Bologna, Italy (in Italian)
21. Metcalf E (2003) Wastewater engineering. Treatment and reuse, 4th edn. McGraw Hill, New York
22. Smith JA, Carliell-Marquet CM (2008) The digestibility of iron-dosed activated sludge. Biores Technol 99:8585–8592
23. Goi D, Tubaro F, Dolcetti G (2006) Analysis of metals and EOX in sludge from municipal wastewater treatment plants: a case study. Waste Manag 26:167–175
24. Austermann-Haun U, Lange R, Seyfried CF, Rosenwinkel KH (1998) Upgrading an anaerobic/aerobic wastewater treatment plant. Water Sci Technol 37(9):243–250
25. Moeller-Chávez G, Seguí-Amórtegui L, Alfranca-Burriel O, Escalante-Estrada V, Pozo-Román F, Rivas-Hernández A (2004) Water reuse in the Apatlaco River Basin (México): a feasibility study. Water Sci Technol 50(2):329–337
26. Tziakis I, Pachiadakis I, Moraitakis M, Xideas K, Theologis G, Tsagarakis KP (2009) Valuing benefits from wastewater treatment and reuse using contingent valuation methodology. Desalination 237:117–125
27. Yang H, Abbaspour KC (2006) Analysis of wastewater reuse potential in Beijing. Desalination 212:238–250

Chapter 2
Detection of Transformation Products of Emerging Contaminants

Anastasia Nikolaou and Giusy Lofrano

Abstract Environmental research nowadays is focusing more and more on different categories of emerging contaminants, because of their increasing release into the environment, the lack of information regarding the occurrence and health effects of the parent compounds, and, even more, of their transformation products, and the need for the development and optimisation of analytical methods for their determination in environmental samples. Emerging contaminants and transformation products are detected by advanced analytical methods such as liquid chromatography (LC) or gas chromatography (GC) combined with tandem mass spectrometric (MS/MS) detection. The rapid development and improvement of these methods during the last few years provides the opportunity not only to determine trace levels of emerging pollutants in environmental samples, but also to identify and detect their transformation products. This is a particularly important step towards safeguarding environmental quality and human health.

Keywords Emerging pollutants · Transformation products · Pharmaceuticals · Personal care products · Environmental samples · LC–MS · GC–MS

A. Nikolaou (✉)
Department of Marine Sciences, University of the Aegean,
University Hill, 81100 Mytilene, Lesvos, Greece
e-mail: nnikol@aegean.gr

G. Lofrano
Department of Civil Engineering, University of Salerno,
via Ponte don Melillo 1, 84084 Fisciano (SA), Italy
e-mail: glofrano@unisa.it

G. Lofrano (ed.), *Green Technologies for Wastewater Treatment*,
SpringerBriefs in Green Chemistry for Sustainability,
DOI: 10.1007/978-94-007-1430-4_2, © Nikolaou, Lofrano 2012

2.1 Introduction

Emerging contaminants are increasingly being released into the environment, raising concerns for environmental quality and human health. These compounds, belonging to several categories, have not been regulated yet. The reason can be either the lack of information regarding their occurrence and environmental effects, or the lack of appropriate analytical methods for their determination in complex environmental samples, or both. Many emerging contaminants are produced and used in major sectors of human life, such as industry, agriculture, consumer goods. Others are unintentionally formed as by-products of industrial processes, advanced oxidation processes, degradation and transformation processes of the original compounds in general. The latter category is gaining more and more scientific interest, as very little is known about these compounds. With the analytical techniques available some years ago, they could not even be identified or detected in environmental samples. However nowadays this has become technically feasible, and therefore much relevant research is being carried out, in order to obtain knowledge about them, and, if necessary, take the required measures to protect the environment and human life.

Emerging contaminants are currently not subject to routine monitoring and may be candidates for future regulation, depending on research on their toxicity and potential health effects. Their inclusion in routine monitoring programs requires analytical methods for their determination in environmental samples to be developed, evaluated and optimised. This can be a difficult task, especially for pollutants occurring in trace levels in environmental samples rich in organic compounds that can interfere during analysis.

The main categories of emerging contaminants are pharmaceuticals, steroid estrogens and personal care products. More categories as well as more compounds and transformation products are being detected in trace levels in the environment, with the help of newly available technology enabling analysts to accurately identify and quantify them in environmental samples. Obtaining information on the sources, occurrence, fate, risk assessment and ecotoxicological data of emerging pollutants is the aim of many investigations worldwide. For this purpose, development and validation of analytical methods appropriate for the determination of these categories of pollutants is a fundamental issue, and the subject of many relevant publications nowadays [1, 2].

The present chapter provides an overview of the recent research findings regarding the determination of the main categories of emerging contaminants and their transformation products in environmental samples.

2.2 Detection of Emerging Contaminants and Their Transformation Products in Environmental Samples

2.2.1 Pharmaceuticals and Transformation Products

Pharmaceuticals consist one of the largest and most important categories of emerging pollutants. During the recent years, several analytical methods have been

Table 2.1 Transformation products of pharmaceuticals identified in environmental samples

Pharmaceuticals	Transformation products	References
Sulfamethoxazole	Hydroxylamine	[3]
Paracetamol	N-4-hydroxyphenyl-acetamide 2-[(2,6-dichlorophenyl)-amino]- 5-hydroxyphenylacetic acid 2,5-dihydroxyphenylacetic acid	[4, 5]
Sulfadiazine	4-methyl-2-amino-pyrimidine	[6, 7]
Sulfamethoxine	2,6-dimethoxy-4-aminopyrimidine 2-aminothiazole	[6, 7]
Sulfathiazole	2,6-dimethoxy-4-aminopyrimidine 2-aminothiazole	[6, 7]
Sulfamerazine	4-methyl-2-aminopyrimidine	[6, 7]
Busperidone	Hydroxybusperidone Dihydroxybusperidone Dipyrimidinylbusperidone 1-pyrimidinyl piperazine	[6, 7]
Carbamazepine	1-(2-benzaldehyde)-4-hydro(1H,3H)quinazoline-2-one 1-(2-benzaldehyde)-(1H,3H)quinazoline-2,4-dione 1-(2-benzoic acid)-(1H,3H)quinazoline-2,4-dione Acridine, salicylic acid, catechol, anthranilic acid	[8, 9]
Iopromide	Primary alcohols Carboxylates	[10]
Acetyl salicylic acid	Salicylic acid	[10]
Clofibrate	Clofibric acid	[10]
Fluoxetine	Norfluoxetine	[10]

developed for the analysis of different groups of pharmaceuticals, because of their different properties. The trend nowadays is to achieve the development and application of "multi-residue" methods for the simultaneous analysis of a large number of pharmaceuticals belonging to different categories [11, 12]. Multi-residue methods have the advantage of providing data on the occurrence of pharmaceuticals in the environment with less time and effort spent. However, the simultaneous analysis of compounds with different physico-chemical properties requires a compromise in regard to the experimental conditions of the analytical method, to accurately determine all analytes in a single run. The same methods used for the determination of the parent compounds are being applied in order to identify and quantify the transformation products of pharmaceuticals as well, based on the MS techniques (Table 2.1).

2.2.2 Detection Methods

During the application of the multi-residue methods, simultaneous extraction of all target analytes from the sample is performed, typically in one single solid phase extraction (SPE) step [13, 14]. Combination of two SPE materials can be performed either in series or classifying the analytes into two or more groups, according to their physico–chemical properties. The most commonly used cartridges for this purpose are Oasis HLB or C18. The value of pH can be very important for this step. For example, for the Oasis HLB neutral sample pH is advisable, whereas for C18 sample pH adjustment prior to extraction is required depending on the acidic, neutral or basic nature of the analytes. Other cartridges used are Lichrolut ENV+ , Oasis MCX and StrataX [2]. The elution is performed with pure organic solvents, mostly methanol or acetonitrile [14–16]. Other extraction methods include Molecularly Imprinted Polymers (MIPs) and immuno-sorbents. They provide high selectivity for target analytes when performing single group analysis, for this reason they have been widely employed to selectively isolate clenbuterol, aniline β-agonists, tetracycline and sulphonamide antibiotics, β-agonists and β-antagonists from biological samples. Some applications also in environmental samples have been reported [17, 18].

After extraction, a purification step is required in order to avoid matrix effects, especially in complex environmental samples. Purification (cleanup) is generally performed by SPE, using the same cartridges and conditions as the analysis of pharmaceuticals in water samples. Sample extracts are therefore diluted with an appropriate volume of MilliQ water, until the organic solvent content is below 10%, in order to avoid losses of target compounds during SPE [19].

LC–MS/MS is the preferable analysis method, due to its versatility, specificity and selectivity, gradually replacing GC–MS and LC–MS. GC–MS can still be successfully applied in some cases, especially for nonpolar and volatile pharma-ceutical compounds, however it still requires a time-consuming derivatization step, during which there are risks of analyte losses [20, 21]. Among LCMS/MS techniques, triple quadrupole (QqQ) and ion trap (IT) instruments are in common use, and they permit the detection of pharmaceuticals at the ng/L range. More recent approaches in LC–MS/MS are linear ion traps (LITs), new generation triple quadrupoles and hybrid instruments, such as quadrupole–time of flight (QqTOF) and quadrupole–linear ion trap (QqLIT) [22]. QqTOF instruments have been used for the elucidation of structures proposed for transformation products [23–25]. They are also used for confirmation purposes. QqLIT methods have also been developed, for the determination of diclofenac, carbamazepine and iodinated X-ray contrast media [26], and for determination of β-blockers in wastewater [27]. Reversed-phase LC is mainly used, with C18 columns. For acidic drugs, ionpair reversed-phase LC with a Phenyl–Hexyl column has also been used [28]. The mobile phases mostly used are acetonitrile, methanol or mixtures. The sensitivity of the method can be improved with use of mobile phase modifiers, buffers and acids, usually ammonium acetate, tri-n-butylamine (TrBA), formic acid and acetic acid [13].

LC–MS/MS methods have been recently used for the determination of the occurrence of illicit drugs and metabolites in water and wastewater with interesting results. Cocainics, amphetamine-like compounds, opiates, cannabinoids and lyser-gics have been determined [14, 29, 30]. The highest levels were reported for cocaine and its main metabolite benzoylecgonine (BE), sometimes reaching the µg/L [14].

Another product of cocaine, cocaethylene (CE), was detected. CE is a transesterification product formed when cocaine is consumed together with ethanol. CE transforms rapidly into metabolites not studied yet in WWTPs, such as norcocaethylene and ecgonine ethyl ester. The cocaine metabolites norcocaine and norbenzoylecgonine, have been determined at two WWTPs in Italy with maximum concentration 40 ng/L. Morphine has been found in some WWTPs at high ng/L levels and heroine at very low concentrations due to its low consumption and its also rapid hydrolysis to morphine and 6-acetylmorphine (heroine is quite unstable in blood serum). Research in WWTPs in Italy and in Switzerland reported that methadone was detected at lower ng/L levels than its pharmacologic inactive metabolite 2-ethylidine-1,5-dimethyl-3,3-diphenylpyrrolidine perchlorate (EDDP). Lysergic acid diethylamide (LSD) and its metabolites nor-LSD and nor-iso LSD (nor-LSD) and 2-oxo-3-hydroxy-LSD (O–H-LSD), have been detected at very low concentrations [14]. Phenylethylamine ephedrine, 3,4-methylenedioxymetamphetamine hydrochloride (MDMA or "ecstasy"), methylenedioxyethylamphetamine (MDE, MDEA or "Eve") and 3,4-methylenedioxyamphetamine (MDA or "Love pills", and metabolite of both MDE and MDMA), have been detected frequently at the ng/L level. 11-nor-9 carboxy THC (nor-THC) and 11-hydroxy-THC (OH-THC), both metabolites of Δ9-tetrahydrocannabinol (THC), which is the most psychologically active constituent of Cannabis have also been detected [30, 31].

2.3 Steroid Estrogens and Transformation Products

Steroid estrogens (hormones and contraceptives) include free estrogens, both natural (estradiol, estrone and estriol) and synthetic (basically ethynyl estradiol, mestranol and diethylstilberol). These compounds have been investigated in environmental samples more than conjugated estrogens and halogenated derivatives recently identified, again by advanced LC–MS and GC–MS techniques [32–35] (Table 2.2).

2.3.1 Detection Methods

SPE is the preferable method for the extraction of estrogens from water samples using mostly cartridges, C18-bonded silica, polymeric graphitized carbon black (GCB) and Oasis HLB [36]. On-line SPE has also been used [37]. Methanol is mainly used for the elution. Use of MIPs for the extraction has also been reported [38], as well as SPME (fibre and in-tube SPME), in combination with either LC or GC instruments [39, 40].

GC–MS and GC–MS/MS have been applied for the determination of estrogens, but their disadvantage is the need for derivatization prior to analysis [39]. Moreover, these methodologies are mainly based on the determination of free estrogens,

Table 2.2 Transformation products of steroid estrogens identified in environmental samples

Steroid estrogens	Transformation products	References
17β-estradiol	Estrone	[32]
	Estriol	[34]
Testosterone	Androstenedione	[34]
Ethinylestradiol (EE2)	2-Cl-EE2	[33]
	2-Br-EE2	[35]
	4-Cl-EE2	
	4-Br-EE2	
	2,4-diCl-EE2	
	2,4-diBr-EE2	

unless intermediate hydrolysis steps are performed [41]. Therefore, LCMS and especially LC–MS/MS are mostly being used [42], which allow the determination of both conjugated and free estrogens without derivatization and hydrolysis. In the case of GC–MS, derivatization is generally carried out in the –OH groups of the steroid ring, performed by silylation with reagents such as N,O-bis(trimethyl-silyl)- acetamide (BSA), N-methyl-N trimethylsilyltrifluoroacetamide (MSTFA), NO-bis(trimethylsilyl)-trifluoroacetamide (BSTFA), or N-(tert-butyldimethylsilyl)-N methyltrifluoroacetamide (MTBSTFA), which lead to the formation of trimethylsilyl (TMS) and tert-butyldimethylsilyl (TBS) derivatives [43]. In the case of LC, octadecyl silica stationary phases are used, with mobile phases consisting of mixtures of water/methanol and, more frequently, water/acetonitrile, sometimes with added modifiers such as 0.1% acetic acid, 0.2% formic acid or 20 mM ammonium acetate. Single and triple quadrupole analysers have been the most widely used for the analysis of estrogens, and application of Q-TOF has been reported as well [44].

Estrogens are mainly excreted as their less active sulphate, glucuronide and sulfo-glucuronide conjugates. However, in the environment, these conjugates may suffer deconjugation and act as precursors of the corresponding free steroids [45]. Thus, an appropriate evaluation of their occurrence and impact requires the analysis of both free and conjugated estrogens. Concentrations reported in wastewater have been most usually in the ng/L range. Estradiol (E2) and estrone (E1) have been the free estrogens most frequently found, whereas estriol (E3) has been studied and detected only sporadically. However, E3 concentrations, when detected, have been usually higher than those of E2 and E1. In general, estrogens concentrations decrease in the order E3 > E1 > E2 [45, 46]. The most studied synthetic estrogen, ethinylestradiol (EE2), has been either not detected [47] or detected at concentrations in general much lower than the other estrogens [48]. Conjugated estrogen derivatives have been less investigated than free ones [49]. Sulphates and glucuronides of E1, E2 and E3 have been determined at similar levels as the free estrogens. Derivatives of the synthetic estrogen EE2 were studied, but they were not detected [50]. From the Di-conjugated E2 derivatives, high levels of the disulphate and moderately high levels of the sulphate-glucuronide derivative were reported [51].

2.4 Personal Care Products and Transformation Products

Personal Care Products (PCPs) are a group of emerging pollutants of particular environmental interest as they are used daily and released into the environment. Synthetic musk fragrances (nitro and polycyclic musk fragrances), antimicrobials (triclosan and its metabolites and triclocarban), sunscreen agents (ultraviolet filters), insect repellents (N,N diethyl-m-toluamide, known as DEET) and parabens (p-hydroxybenzoic esters), which are basically substances used in soaps, shampoos, deodorants, lotions, toothpaste and other PCPs are some of the major substances of this group. The nitro musk fragrances were the first to be produced and include musk xylene, ketone, ambrette, moskene and tibetene. In the environment, the nitro substituents can be reduced to form amino metabolites of these compounds. The polycyclic musk fragrances, which are used in higher quantities than nitro musks, include 1,2,4,6,7,8-hexahydro-4,6,6,7,8,8-hexamethylcyclopenta-γ-2-benzopyrane (HHCB), 7-acetyl-1,1,3,4,4,6-hexamethyl-1,2,3,4-tetrahydronaphthalene (AHTN), 4-acetyl-1, 1-dimethyl-6-tert-butylindane (ADBI), 6-acetyl-1,1,2,3,3,5-hexamethylindane (AHMI), 5-acetyl-1,1,2,6-tetramethyl-3-isopropylindane (ATII) and 6,7-dihydro-1,1,2,3,3-penta-methyl-4-(5H)-indanone (DPMI). Parabens are the most common preservatives used in personal care products and in pharmaceuticals and food products. This group of substances includes methylparaben, propylparaben, ethylparaben, butylparaben and benzylparaben. Transformation products of PCPs have also been detected in environmental samples (Table 2.3).

2.4.1 Detection Methods

PCPs and transformation products are extracted from the matrix by liquid–liquid extraction (LLE) [52], continuous liquid–liquid extraction (CLLE), SPE [53] and SPME [54]. LLE and CLLE have been applied with various organic solvents such as dichloromethane, pentane, hexane, toluene, cyclohexane and petroleum ether, or mixtures of them. For SPE, sorbents commonly used are C18, Isolute ENV+, OasisMAX, SDB-XC, XAD-2, and XAD-4/XAD-8. Several organic solvents have been used for the elution, such as acetone, methanol, toluene, hexane, ethyl acetate and their mixtures. Cleanup is performed with SPE with silica and alumina [53].

The analysis of PCPs is performed mainly by GC–MS techniques, in particular by GC–EI–MS or GC–NCI–MS. The latter is more sensitive for the category of nitro musk fragrances. These compounds have also been analysed by GC–FID, GC–ECD, and high-resolution and ion trap tandem mass spectrometry (MS/MS). Triclosan and its chlorinated metabolites are also determined by GC–EI–MS with and without derivatization, LC–MS and LC–MS/MS. When derivatizing, N,N-diethyltrimethylamine (TMS-DEA), (BSTFA), pentafluorinated triclosan and tert-butyldimethylsilyl triclosan are the ether derivatives generated after reaction with methyl chloroformate (MCF), pentafluoropropionic acid anhydride (PFA) and

Table 2.3 Transformation products of PCPs identified in environmental samples

PCPs	Transformation products	References
Paraben	4-hydroxybenzoic acid Phenol	[55]
Triclosan	Methyl-triclosan	[56]
Nitroaromatic musks	Aniline transformation products	[57]
HHCB	HHCB lactone	[57]

Ntert-butyldimethylsilyl-N-methyltrifluoroacetamide (MTBSTFA), respectively. UV filters and insect repellants are determined mainly with GC–MS analysis. Almost all UV filters are amenable to GC except octyl triazone, avobenzone, 4-isopropyldibenzoylmethane and 2-phenylbenzimidazole-5-sulphonic acid, that can be determined by HPLC–UV. Parabens are analysed by LC–MS/MS methods [37].

2.5 Concluding Remarks

The determination of emerging contaminants and of their transformation products in complex environmental samples is a challenging task. These compounds occur in the environment in trace concentrations, however due to their continuous release they can have adverse effects of aquatic organisms and the trophic chain. Moreover, their oxidation/degradation products and metabolites have not yet been fully documented, but with advanced analytical techniques currently available they can be identified and quantified, providing more insight to the occurrence, formation, properties and environmental pathways of these pollutants.

References

1. Fatta D, Achilleos A, Nikolaou A, Meric S (2007) Analytical methods for tracing pharmaceutical residues in water and wastewater. TrAC-Trends Anal Chem 26(6):515–533
2. Kostopoulou M, Nikolaou A (2008) Analytical problems and the need for sample preparation in the determination of pharmaceuticals and their metabolites in aqueous environmental matrices. TrAC-Trends Anal Chem 27(11):1023–1035
3. Munoz F, von Sonntag C (2000) The reactions of ozone with tertiary amines including the complexing agents nitrilotriacetic acid. Chem Soc Perkin Trans 2:2029–2037
4. Vogna D, Marotta R, Napolitano A, d'Ischia M (2002) UV/H2O2-induced hydroxylation/ degradation pathways and 15 N-aided inventory of nitrogenous breakdown products. J Org Chem 67(17):6143–6151
5. Andreozzi R, Caprio V, Marotta R, Vogna D (2003) Paracetamol oxidation from aqueous solutions by means of ozonation and H_2O_2/UV system. Water Res 37:993–999
6. Calza P, Pazzi M, Medana C, Baiocchi C, Pelizzetti E (2004) The photocatalytic process as a tool to identify metabolitic products formed from dopant substances: the case of buspirone. J Pharm Biomed Anal 35:9–19

7. Calza P, Medana C, Pazzi M, Baiocchi C, Pelizzetti E (2004) Photocatalytic transformations of aminopyrimidines on TiO$_2$ in aqueous solution. Appl Catal B Environ 52:267–274
8. Snyder SA, Wert EC, Rexing DJ, Zegers RE, Drury DD (2006) Ozone oxidation of endocrine disruptors and pharmaceuticals in surface water. Ozone Sci Eng 28:445–453
9. Vogna D, Marotta R, Napolitano A, Andreozzi R, d'Ischia M (2004) Advanced oxidation of the pharmaceutical drug diclofenac with UV/H2O2 and ozone. Water Res 38(2):414–422
10. Farré M, Péreza S, Kantiania L, Barceló D (2008) Fate and toxicity of emerging pollutants, their metabolites and transformation products in the aquatic environment. TrAC-Trends Anal Chem 27(11):991–1007
11. Botitsi E, Frosyni C, Tsipi D (2007) Determination of pharmaceuticals from different therapeutic classes in wastewaters by liquid chromatography–electrospray ionization–tandem mass spectrometry. Anal Bioanal Chem 387:1317–1327
12. Gros M, Petrovic M, Barcelo D (2006) Development of a multi-residue analytical methodology based on liquid chromatography–tandem mass spectrometry (LC–MS/MS) for screening and trace level determination of pharmaceuticals in surface and wastewaters. Talanta 70:678–690
13. Gros M, Petrovic M, Barcelo D (2006) Multi-residue analytical methods using LC-tandem MS for the determination of pharmaceuticals in environmental and wastewater samples: a review. Anal Bioanal Chem 386:941–952
14. Petrovic M, Gros M, Barcelo D (2006) Multi-residue analysis of pharmaceuticals in wastewater by ultra-performance liquid chromatography-quadrupole-time-of-flight-mass spectrometry. J Chromatogr A 1124:68–81
15. Bryan PD, Hawkins KR, Stewart JT, Capomacchia AC (1992) Analysis of chlortetracycline by high performance liquid chromatography with postcolumn alkaline-induced fluorescence detection. Biomed Chromatogr 6:305–310
16. Diaz-Cruz MS, Barcelo D (2006) Determination of antimicrobial residues and metabolites in the aquatic environment by liquid chromatography tandem mass spectrometry. Anal Bioanal Chem 386:973–985
17. Bravo JC, Garcinuno RM, Fernandez P, Durand JS (2007) A new molecularly imprinted polymer for the on-column solid-phase extraction of diethylstilbestrol from aqueous samples. Anal Bioanal Chem 388:1039–1045
18. O'Connor S, Aga DS (2007) Analysis of tetracycline antibiotics in soil: advances in extraction, clean-up, and quantification. TrAC-Trends Anal Chem 26:456–465
19. Jacobsen AM, Halling-Sorensen B, Ingerslev F, Hansen SH (2004) Simultaneous extraction of tetracycline, macrolide and sulfonamide antibiotics from agricultural soils using presurrised liqiod extraction followed by soild-phase extraction and liquid chromatography-tandem mass spectrometry. J Chromatogr A 1038:157–170
20. Kolpin DW, Furlong ET, Meyer MT, Thurman EM, Zaugg SD, Barber LB, Buxton HT (2002) Pharmaceuticals, hormones, and other organic wastewater contaminants in U.S. streams, 1999–2000—A national reconnaissance. Environ Sci Technol 36:1202–1211
21. Weigel S, Berger U, Jensen E, Kallenborn R, Thoresen H, Huhnerfuss H (2004) Determination of selected pharmaceuticals and caffeine in sewage and seawater from Tromsø/Norway with emphasis on ibuprofen and its metabolites. Chemosphere 56:583–592
22. Perez S, Barcelo D (2007) Application of advanced MS techniques to analysis and identification of human and microbial metabolites of pharmaceuticals in the aquatic environment. TrAC-Trends Anal Chem. 26:494–514
23. Eichhorn P, Ferguson PL, Perez S, Aga DS (2005) Application of ion trap-MS with H/D exchange and QqTOF-MS in the identification of microbial degradates of trimethoprim in nitrifying activated sludge. Anal Chem 77:4176–4184
24. Gomez MJ, Malato O, Ferrer I, Aguera A, Fernandez-Alba AR (2007) Solid-phase extraction followed by liquid chromatography—time-of-flight—mass spectrometry to evaluate pharmaceuticals in effluents. A pilot monitoring study. J Environ Monit 9:718–729
25. Stolker AAM, Niesing W, Hogendoorn EA, Versteegh JFM, Fuchs R, Brinkman UAT (2004) Liquid chromatography with triple-quadrupole or quadrupole-time of flight mass

spectrometry for screening and confirmation of residues of pharmaceuticals in water. Anal Bioanal Chem 378(9):955–963

26. Seitz W, Schulz W, Weber WH (2006) Novel applications of highly sensitive liquid chromatography/mass spectrometry/mass spectrometry for the direct detection of ultra-trace levels of contaminants in water. Rapid Commun Mass Spectrom 20:2281–2285

27. Nikolai LN, McClure EL, MacLeod SL, Wong CS (2006) Stereoisomer quantification of the β-blocker drugs atenolol, metoprolol, and propranolol in wastewaters by chiral high performance liquid chromatography-tandem mass spectrometry. J Chromatogr A 1131: 103–109

28. Quintana JB, Reemtsma T (2004) Sensitive determination of acidic drugs and triclosan in surface and wastewater by ion-pair reverse-phase liquid chromatography/tandem mass spectrometry. Rapid Commun Mass Spectrom 18:765–774

29. Hada M, Takino M, Yamagami T, Daishima S, Yamaguchi K (2000) Trace analysis of pesticide residues in water by high-speed narrow-bore capillary gas chromatography–mass spectrometry with programmable temperature. J Chromatogr A 874:81–90

30. Santos FJ, Galceran MT (2002) The application of gas chromatography to environmental analysis. TrAC-Trends Anal Chem 21:672–685

31. Stuber M, Reemtsma T (2004) Evaluation of three calibration methods to compensate matrix effects in environmental analysis with LC-ESI-MS. Anal Bioanal Chem 378:910–918

32. Ferguson PL, Iden CR, McElroy AE, Brownawell BJ (2001) Determination of steroid estrogens in wastewater by immunoaffinity extraction coupled with HPLC-electrospray-MS. Anal Chem 73:3890–3895

33. Kuster M, Lopez de Alda MJ, Barcelo D (2004) Analysis and distribution of estrogens and progestogens in sewage sludge, soils and sediments. TrAC-Trends Anal Chem 23:790–798

34. Lee LS, Strock TJ, Sarmah AK, Rao PSC (2003) Sorption and Dissipation of Testosterone, Estrogens, and Their Primary Transformation Products in Soils and Sediment. Environ Sci Technol 37(18):4098–4105

35. Lee Y, Von Gunten U (2007) Efficient removal of the estrogenic activity during oxidative treatment of waters containing steroid estrogens, Conference Micropol and Ecohazard 2007, Frankfurt

36. Fine DD, Breidenbach GP, Price TL, Hutchins SR (2003) Quantitation of estrogens in ground water and swine lagoon samples using solid-phase extraction, pentafluorobenzyl/ trimethylsilyl derivatizations and gas chromatography–negative ion chemical ionization tandem mass spectrometry. J Chromatogr A 1017:167

37. Rodriguez-Mozaz S, Lopez de Alda MJ, Barcelo D (2004) Picogram per liter level determination of estrogens in natural waters and waterworks by a fully automated on-line solid-phase extraction-liquid chromatography-electrospray tandem mass spectrometry method. Anal Chem 76:6998–7006

38. Watabe Y, Kubo T, Nishikawa T, Fujita T, Kaya K, Hosoya K (2006) Fully automated liquid chromatography–mass spectrometry determination of 17β-estradiol in river water. J Chromatogr A 1120:252–259

39. Mitani K, Fujioka M, Kataoka H (2005) Fully automated analysis of estrogens in environmental waters by in-tube solid-phase microextraction coupled with liquid chromatography-tandem mass spectrometry. J Chromatogr A 1081:218–224

40. Penalver A, Pocurull E, Borrull F, Marce RM (2002) Method based on solid-phase microextraction–high-performance liquid chromatography with UV and electrochemical detection to determine estrogenic compounds in water samples. J Chromatogr A 964:153–160

41. Isobe T, Shiraishi H, Yasuda M, Shinoda A, Suzuki H, Morita M (2003) Determination of estrogens and their conjugates in water using solid-phase extraction followed by liquid chromatography-tandem mass spectrometry. J Chromatogr A 984:195–202

42. Lopez de Alda MJ, Diaz-Cruz S, Petrovic M, Barcelo D (2003) Liquid chromatography-(tandem) mass spectrometry of selected emerging pollutants (steroid sex hormones, drugs and alkylphenolic surfactants) in the aquatic environment. J Chromatogr A 1000:503–526

43. Shareef A, Angove MJ, Wells JD (2006) Optimization of silylation using N -methyl- N -(trimethylsilyl)-trifluoracetimide, N, O -bis-(trimethylsilyl)-trifluoroacetamide and N -(tert -butyldimethylsilyl)- N -methyltrifluoroacetamide for the determination of the estrogens estrone and 17 a -ethinylestradiol by gas chromatography-mass spectrometry. J Chromatogr A 1108:121–128

44. Reddy S, Brownawell BJ (2005) Analysis of estrogens in sediment from a sewage-impacted urban estuary using high-performance liquid chromatography/time-of-flight mass spectrometry. Environ Toxicol Chem 24:1041–1047

45. Lamprecht G, Kraushofer T, Stoschitzky K, Lindner W (2000) Enantioselective analysis of (R)- and (S)-atenolol in urine samples by a high-performance liquid chromatography column-switching setup. J Chromatogr B Biomed Sci Appl 740:219–226

46. Vrana B, Allan IJ, Greenwood R, Mills GA, Dominiak E, Svensson K, Knutsson J, Morrison G (2005) Passive sampling techniques for monitoring pollutants in water. TrAC-Trends Anal Chem 24:845–868

47. Bruheim I, Liu X, Pawliszyn J (2003) Thin film microextraction. Anal Chem 75:1002–1010

48. Rodriguez I, Rubi E, Gonzalez R, Quintana JB, Cela R (2005) On-fibre silylation following solid-phase microextraction for the determination of acidic herbicides in water samples by gas chromatography. Anal Chim Acta 537:259–266

49. Koester CJ, Simonich SL, Esser BK (2003) Environmental analysis. Anal Chem 75:2813–2829

50. Lord H, Pawliszyn J (2000) Evolution of solid-phase microextraction technology. J Chromatogr A 885:153–193

51. Wu J, Yu X, Lord H, Pawliszyn J (2000) Solid-phase microextraction of inorganic ions based on polypyrrole film. Analyst 125:391–394

52. Pang X, Cheng G, Li R, Lu S, Zhang Y (2005) Bovine serum albumin-imprinted polyacrylamide gel beads prepared via inverse-phase seed suspension polymerization. Anal Chim Acta 550:13–17

53. Peck AM (2006) Analytical methods for the determination of persistent ingredients of personal care products in environmental matrices. Anal Bioanal Chem 386:907–916

54. Artola-Garicano E, Borkent I, Hermens JLM, Vaes WHJ (2003) Removal of two polycyclic musks in sewage treatment plants: freely dissolved and total concentrations. Environ Sci Technol 37:3111–3116

55. Valkova N, Lépine F, Valeanu L, Dupont M, Labrie L, Bisaillon JG, Beaudet R, Shareck F, Villemur R (2001) Hydrolysis of 4-hydroxybenzoic acid esters (Parabens) and their aerobic transformation into phenol by the resistant enterobacter cloacae strain EM. Appl Environ Microbiol 67(6):2404–2409

56. Balmer ME, Poiger T, Droz C, Romanin K, Bergqvist PA, Müller MD, Buser HR (2004) Occurrence of methyl triclosan, a transformation product of the bactericide triclosan, in fish from various lakes in Switzerland. Environ Sci Technol 38(2):390–395

57. Bester K (2009) Analysis of musk fragrances in environmental samples. J Chromatogr A 1216(3):470–480

Chapter 3
Removal of Trace Pollutants by Application of MBR Technology for Wastewater Treatment

Giorgio Bertanza and Roberta Pedrazzani

Abstract The number of MBR technology applications for municipal wastewater treatment has grown enormously during the last decade. These systems have several environmental advantages with respect to conventional activated sludge process (high effluent quality, reduced footprint, reduced sludge production etc.); recently, it has been demonstrated that they improve at higher extent effluent estrogenicity abatement, even though this positive effect may not be predicted based on trace pollutant concentrations, being for many substances similar to those obtained by conventional process. On the other hand, higher environmental costs, mainly due to membrane fabrication/installation and energy consumption for plant operation, have to be carefully evaluated in order to assess if MBR can be considered to be a green technology.

Keywords Bioassays · EDCs · Efficiency · MBR · Process parameters

3.1 Introduction

Biological wastewater treatment can be considered a "green" technology *a priori*: actually, it exploits natural processes which are somehow controlled.

Microorganism metabolism is responsible for contaminants either transformation into innocuous compounds (e.g., mineralization of organic carbon to CO_2 and H_2O,

G. Bertanza (✉) · R. Pedrazzani
Faculty of Engineering, University of Brescia,
via Branze 43, 25123 Brescia, Italy
e-mail: bert@ing.unibs.it

R. Pedrazzani
e-mail: roberta.pedrazzani@ing.unibs.it

G. Lofrano (ed.), *Green Technologies for Wastewater Treatment*,
SpringerBriefs in Green Chemistry for Sustainability,
DOI: 10.1007/978-94-007-1430-4_3, © Bertanza, Pedrazzani 2012

oxidation of ammonia to nitrate and subsequent reduction to molecular nitrogen), or transfer of pollutants from the liquid to the solid phase (sorption onto the biomass of highly hydrophobic pollutants and colloids, cell assimilation of nutrients and carbon) so that they can be removed via the excess sludge disposal.

In order to make these processes suitable for human needs, an engineered environment is required: the so called Waste Water Treatment Plant (WWTP). Activated sludge process is the most widely used in the world. It comprises several technical devices and equipments such as: concrete tanks, pipes, valves, blowers, air diffusers, sensors, control devices etc. Nevertheless, this machinery is aimed basically at quickening and regulating natural processes.

Membrane Bio Reactor (MBR) technology does not differ, in principle, by an activated sludge WWTP. In effect, a conventional activated sludge plant can be transformed into an MBR system by adding a membrane filtration unit in place of the final sedimentation tank. This means that a very simple treatment phase (sedimentation) is replaced by a technological device (membrane), with possible higher capital and operating costs. Obviously, this can be accepted in case valuable advantages are provided by MBR technology. Actually, some of them are well known: reduced plant footprint, high quality effluent (as far as suspended solids related contaminants are concerned), pathogens removal capacity, etc.

The challenge, nowadays, is to assess if and at which extent MBR technology is helpful in reducing (with respect to conventional activated sludge process) trace pollutants discharge into the environment.

In this chapter, a brief description of MBR principles and technology is reported, with updated figures about the diffusion of these plants worldwide. The efficiency of MBR plants in trace pollutant removal is then shown, based on literature data, in comparison with conventional systems. Finally, recently published results are used to discuss upon the "green" character of MBR technology.

3.2 The MBR Technology

Membrane Bio Reactor (MBR) is the acronym for water and wastewater treatment processes which integrates a biological process with a membrane separation step. In general, membrane filtration is aimed at retaining biomass and other suspended materials so as to produce a clarified and disinfected permeate.

First applications at full scale of MBRs were developed for treating ship-board wastewater in the late 1960s [1, 2]. Relevant commercial developments took place from the late 1980s to the early 1990s, in Japan, pushed by government programmes, which promoted, among the others, the research activity of the agricultural machinery company, Kubota. As a consequence, by the end of 1996, 60 Kubota plants were in operation in Japan for domestic wastewater treatment, and later on for industrial effluent treatment, for a total capacity of 5.5 thousands of cubic meters per day [3]. In the same period, in the USA, Zenon Environmental was working on this project and by the early 1990s, the first patent

Table 3.1 MBR membrane module products, municipal market—Reprinted from Ref. [3] with kind permission of Elsevier

Supplier	Country	Date launched	Acquired	Date, first >10 MLD plant[a]
Asahi Kasei	Japan	2004	–	2007
GE-*ZeeWeed*®	USA	1993	Jun-06	2002
Korea Membrane Separation -*KSMBR*®	Korea	2000	–	2008
Koch Membrane System -*PURON*®	USA	2001	Nov-04	2010[b]
Kubota *EK*	Japan	1990	–	1999
Kubota *RW*	Japan	2009	–	–
Memstar	Singapore	2005	–	2010[b]
MICRODYN-NADIR	Germany	2005	–	2010[b]
Mitsubishi Rayon (*SADF*)	Japan	2005	–	2006
Mitsubishi Rayon (*SUR*)	Japan	1993		–
Motimo	China	2000	–	2007
Norit	Netherlands	2002	–	2010[b]
Siemens Water Tech. -*MEMCOR*®	Germany	2002	Jul-04	2008
Toray	Japan	2004	–	2010[b]

[a] MLD = thousands of cubic meters per day
[b] Projected 2010 or 2011

was registered [4, 5]. At the end of the Millennium, 150 thousands of cubic meters per day total capacity of Zenon plants were installed [3].

By the end of the 1990s and the early 2000s, an enormous amount of research and commercial activities led to the development of several MBR membrane products and systems: at least ten membrane products originating from seven Countries were launched. In Table 3.1, the 12 major suppliers on the market in 2010 are listed. However, today dozens of MBR membrane products or technologies are currently commercially available, and MBR systems are in operation in more than 200 Countries, although the municipal wastewater market is still dominated by the original three suppliers Kubota, GE Zenon and Mitsubishi Rayon Engineering: Fig. 3.1 [3].

Membrane configurations used for MBRs are essentially the following: plat-and-frame/flat sheet (FS), hollow fiber (HF), (multi)tubular (MT); flow schemes for the different solutions are shown in Fig. 3.2 [3]. Among all available configurations, these permit adequate turbulence (i.e., cleaning) of membrane surface during operation.

Membrane modules can be submerged or immersed in the biological reactor (iMBR configuration) or sidestream (sMBR configuration). Description of these different configurations is out of the scope of this chapter and is reported in books listed at the end of this paragraph.

Notwithstanding the membrane and MBR process configuration, the effective pore size of filtration is generally below 0.1 μm, so that a clarified and

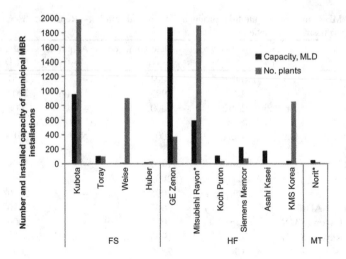

Fig. 3.1 MBR municipal market—Reprinted from Ref. [3] with kind permission of Elsevier; *estimated figures

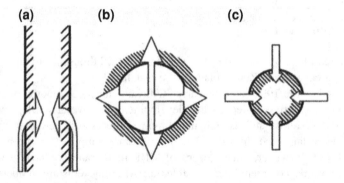

Fig. 3.2 Schematics showing flow through membrane configured as: (**a**) FS, (**b**) MT and (**c**) HF—Reprinted from Ref. [3] with kind permission of Elsevier

substantially disinfected permeate is produced. In addition, suspended solids retainment allows to concentrate the biomass in the biological reactor (reducing required tank volume) and to recirculate sorbed recalcitrant compounds.

Even if in MBR plants sludge settling properties do not affect effluent quality, it has to be underlined that hydraulic and organic shocks can have other serious impacts on the operation of an MBR [3]. Moreover, a critical factor which requires particular care and efforts on design, technology development, operation, is represented by membrane fouling, which was and still is one of the main challenges for MBR processes.

This issue, in effect, extensively dealt with in Judd and Judd [3] and other MBR handbooks, where many other aspects are analyzed in detail as well. Among the others, the following can be cited: *Membrane Bioreactors for Wastewater Treatment* (2000), by T. Stephenson, S. Judd, B. Jefferson and K. Brindle (IWA Publishing); *Membranes for Industrial Wastewater Recycling and Reuse* (2003), by S. Judd and B. Jefferson (Elsevier); *Membrane Systems for Wastewater Treatment* (WEFPress, 2006); *Membrane Technology for Waste Water Treatment* (2006), by J. Pinnekamp and H. Friedrich (FiW-Verlag); *MBR Practice Report: Operating Large Scale Membrane Bioreactors for Municipal Wastewater Treatment* (2011), by C. Brepols (IWA Publishing); *The Guidebook to Membrane Technology for Wastewater Reclamation* (2011), by M. Wilf (Balaban Publishers).

3.3 Trace Pollutants Removal by MBR Plants

Based on inherent features of MBR technology, it may be argued that only particles significantly smaller than the effective membrane pore size and non-biodegradable dissolved compounds can flow through the membranes with the effluent. The affinity of (trace) pollutants for the sludge solids (which are retained inside the reactor) represent a key factor for their degradation/removal. Actually, degradation takes place only if bacterial species capable of metabolizing these compounds can develop and, depending on growth kinetics, SRT (Sludge Retention Time) should play an important role. Nevertheless, literature data are contradictory on the effect of extending the SRT in MBR systems as far as the removal of trace organics is concerned. As reported by Judd and Judd [3], a recent literature review on pharmaceuticals and personal care products (PPCPs) has revealed that [6]:

- readily removed pharmaceuticals are equally well removed by conventional activated sludge process and MBRs; it is the case of acetaminophen, ibuprofen, paroxetine;
- no appreciable difference has been shown also for a few moderately removed species (sotalol and hydrochlorothiazide) and highly intractable species (carbamazepine and some macrolides), whose degradation undergoes several physico-biochemical processes; and
- for other compounds, only slightly better performance is exerted by MBR.

Little appreciable difference between conventional activated sludge and MBR has been also reported by Clara et al. [7] and Cirja et al. [8], but significant improvements have been documented by Bernhard et al. [9] for poorly biodegradable persistent polar compounds (diclofenac, mecoprop, sulfophenylcarboxylates), Kimura et al. [10] for chemically complex substances (ketoprofen and naproxene), Radjenovic et al. [11, 12] for fluoxetine. Authors attributed these results to the long SRT attainable with MBR. As a confirmation of this hypothesis, similar performances were obtained by conventional processes operated at the same long SRTs [7]. However, other scientists found no clear correlation between SRT and removal of trace organics [13–15].

Among other process parameters, hydraulic residence time (HRT) could be expected to affect trace pollutants removal. Actually, Joss et al. [16] evidenced that the biodegradation of some EDCs follows pseudo-first-order kinetics.

An interesting correlation has been reported between pharmaceuticals [17, 18] as well as EDCs [19, 20] removal and nitrification, specific enzymes being postulated to be able to catalyze co-metabolization of several refractory organics.

It follows that MBR has some features which enable this technology to improve (sometimes slightly) the removal of some trace pollutants with respect to conventional activated sludge process. Under the same SRT, temperature and other process conditions (e.g., occurrence of nitrification), these better performances seem to be ascribed to high retention of suspended solid, and, consequently, of specialized slow-growing bacteria and of the organics to be degraded (in case they are adsorbed onto the suspended solids).

3.4 Is MBR a Green Technology?

Factors for which MBR may be considered a green technology are directly related to reduced environmental impact achievable by means of this technology. We can mention in particular the high effluent quality (as already pointed out), the avoided use of chemicals for disinfection (thanks to the capability of membranes to retain pathogens), the reduced excess sludge production (which is a consequence of the generally long SRT), etc. However, as far as trace pollutant removal is concerned, data reported in the previous paragraph do not seem to show significant and general better performances of MBR, in comparison with conventional activated sludge process. At least, this is true if residual concentration of micropollutants released into the environment through the effluent is considered. Nevertheless, what is important is the effective environmental impact of these substances rather then the residual concentration of single compounds and their metabolites. In effect, chemical analysis alone is not useful to investigate synergistic effects among mixtures of different pollutants and their degradation by-products well-known phenomenon in case of endocrine disruptors: [21–23]. Several Authors [24–27] have pointed out that water biological activity should be also monitored, in order to better evaluate treatment suitability; actually, endocrine activity assays have been proposed since the last few years [28–36].

Bertanza et al. [37] proposed an integrated assessment procedure, based on both chemical and biological analyses, to evaluate the performance of biological oxidation in removal of target EDCs from municipal wastewater. The following estrogen-like substances were considered: 4-nonylphenol (NP), its parent compounds 4-nonylphenol monoethoxylate (NP1EO) and 4-nonylphenol diethoxylate (NP2EO), and bisphenol A (BPA). Experimental work was conducted at two full scale WWTPs located in Northern Italy, equipped with either conventional settling tanks (CAS, Conventional Activated Sludge) or with ultrafiltration unit (MBR, Membrane Biological Reactor). Hormonal activity in water samples was

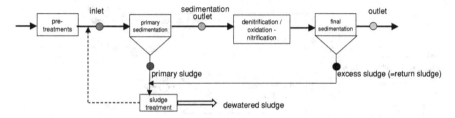

Fig. 3.3 Flowsheet and sampling points for the first (conventional activated sludge) plant (*bold line* = wastewater; *fine line* = sludge; *dotted line* = supernatant from sludge treatment; *double line* = dewatered sludge)—Reprinted from Ref. [37] with kind permission of Elsevier

measured by means of human breast cancer MCF-7 based reporter gene assay, using 17β-estradiol (E2) as a standard.

The studied CAS plant (Fig. 3.3) has a design size of 370,000 p.e. and treats mainly domestic wastewater. The process scheme includes primary settling (volume = 10,400 m³, 3 parallel basins); pre-denitrification (volume = 7,200 m³, 5 parallel basins); oxidation-nitrification (volume = 16,600 m³, 5 parallel basins); secondary settling (volume = 26,100 m³, 6 parallel basins).

The second studied WWTP (Fig. 3.4) consists of 2 CAS lines and 1 MBR line (design size 380,000 p.e.) and treats domestic and industrial wastewater. The process scheme includes equalization/homogenization (volume = 24,000 m³); pre-denitrification (volume = 11,100 m³, 3 parallel basins); oxidation-nitrification (volume = 20,600 m³, 3 parallel basins); secondary settling (for conventional lines, volume = 7,800 m³, 2 parallel basins) and ultrafiltration (for MBR line). This configuration enabled the comparison of the CAS process with the MBR technique.

Removal efficiency and residual effluent concentration of target compounds for studied plants and processes are compared in Fig. 3.5; influent concentrations are reported in Table 3.2.

The experimental results show that, while the CAS and MBR lines of the second plant, where different sludge ages were kept (9 d for CAS and 15 d for MBR, respectively), yielded similar performances, the first CAS plant, having the same sludge age as the MBR line, yielded on the contrary to slightly lower removal efficiencies (apart from BPA). This phenomenon might be due to different sewage temperature (16 °C and 23 °C for the first and second WWTPs, respectively). Actually, it is known that sludge age and temperature are crucial parameters: Clara et al. [38] argue that the minimum required sludge age is 10 d at 10 °C, and further increases do not lead to noticeable improvements. Moreover, in accordance with data reported in the previous paragraph, several Authors [19, 20, 39] conclude that EDC removal occurs only in plants equipped with nitrification stages (as in the studied WWTPs). In addition, Clara et al. [7] report that possible MBR efficiency improvements might be ascribed to an increase in sludge age, rather than to filtration.

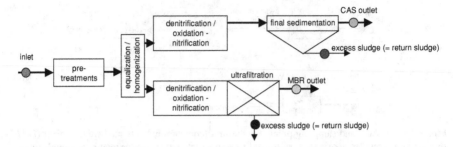

Fig. 3.4 Flowsheet and sampling points for the second (conventional + MBR) plant (*bold line* = wastewater; *fine line* = sludge)—Reprinted from Ref. [37] with kind permission of Elsevier

Nevertheless, biological measurements showed that estrogenic activity (expressed as RLU, Relative Light Units, and normalized toward protein concentration) was reduced to a greater extent by the MBR process with respect to CAS treatment, even if analytes were removed at a comparable level. In effect, water samples (influent and both CAS and MBR effluents of the second WWTP), taken on three different days of consecutive weeks (W1, W2 and W3) during the monitoring period, were submitted to biological assays. Figure 3.6 shows the results of the biological analyses. It is clear that estrogenic activity was significantly reduced by both treatments, and, in five of six cases, with greater efficiency by the MBR system. This is a relevant outcome which emphasizes the importance of biological analyses: actually, while EDC (NP + BPA) concentrations were similar in outlet samples taken from both lines (Fig. 3.6), estrogenic activity exerted by CAS effluent was almost always higher.

The reason of this different performance is still under investigation; likely it might be attributed to metabolic pathways exhibited by different microbial consortia growing in MBR plants [40, 41]. In summary, reported data show that MBR technology can exert actual advantages in terms of trace pollutant removal (at least it was demonstrated concerning the effect of EDCs). Nevertheless it has to be investigated if increased environmental costs (due to membrane and equipment fabrication and additional energy consumption for plant operation) are compensated by reduction in environmental impact obtained with better effluent quality.

3.5 Concluding Remarks

The number of MBR technology applications for municipal wastewater treatment has grown enormously during the last decade, with many new companies competing on the market. This was possible thanks to important results obtained by intense research activity at the end of the past Millennium, which finally led to cost reduction for both manufacturing/installation and operation of MBR plants. In the future, a further diffusion of these systems can be expected, depending on the

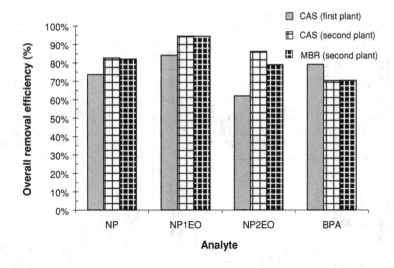

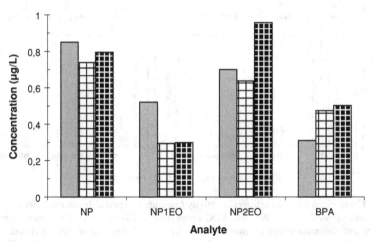

Fig. 3.5 Comparison among studied processes: treatment efficiency (*top*) and effluent residual concentrations (*bottom*) of target EDCs—Reprinted from Ref. [37] with kind permission of Elsevier

Table 3.2 Average concentrations (μg/L) of target EDCs in the influent wastewater for studied WWTPs—Reprinted from Ref. [37] with kind permission of Elsevier

EDC	First WWTP	Second WWTP
NP	4.15	4.70
NP1EO	3.90	7.89
NP2EO	2.18	5.01
BPA	2.19	1.94

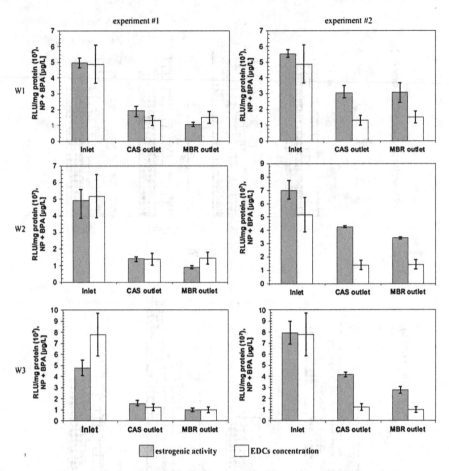

Fig. 3.6 Comparison between estrogenic activity (measured in two different experiments and in three different days W1, W2, W3) and EDC concentration (NP + BPA). *Error bars* represent maximum and minimum values measured in three replicates in the case of biological data, while they show variation percentage in the case of chemical analyses—Reprinted from Ref. [37] with kind permission of Elsevier

strength of driving factors like [3]: legislation, local water shortage, return on investment, environmental impact, public and political acceptance.

MBR system offers a series of environmental advantages, with respect to conventional activated sludge, which may suggest it to be a green technology: reduced plant footprint, high quality effluent (as far as suspended solids related contaminants are concerned), pathogens removal capacity, avoided use of chemicals for disinfection, reduced sludge production, etc.

On the contrary, as far as trace pollutants are concerned, removal of single compounds is often similar to conventional systems, when they are operated under the same operating conditions, in particular in terms of SRT, temperature, HRT, and nitrogen removal processes. Nevertheless, recently it has been demonstrated that the effluent of MBR plants can exert lower estrogenic activity, in comparison with conventional activated sludge, even if no appreciable difference in EDCs concentration is detected.

In summary, it can be concluded that significant environmental properties of MBR technology should be evaluated in relation to increased environmental costs mainly due to membrane and equipment fabrication and additional energy consumption. It is the author's opinion that these evaluations should be carefully performed in the near future.

References

1. Bailey J, Bemberis I, Presti J (1971) Phase I final report—shipboard sewage treatment system. General Dynamics Electric Boat Division NTIS
2. Bemberis I, Hubbard PJ, Leonard FB (1971) Membrane sewage treatment systems potential for complete wastewater treatment. In: American society for agricultural engineering winter meeting, pp 871–878
3. Judd S, Judd C (2011) The MBR book. Principles and applications of memrane bioreactors for water and wastewater treatment, 2nd edn. IWA Publishing, Elsevier, Oxford. ISBN 978-1-84-339518-8
4. Tonelli FA, Behmann H (1996) Aerated membrane bioreactor process for treating recalcitrant compounds. US Pat: 410730
5. Tonelli FA, Canning RP (1993) Membrane bioreactor system for treating synthetic metal-working fluids and oil based products. USA Pat: 5204001
6. Sipma J, Osuna B, Collado N, Monclus H, Ferrero G, Comas J et al (2010) Comparison of removal of pharmaceuticals in MBR and activated sludge systems. Desalination 250:653–659
7. Clara M, Strenn B, Ausserleitner M, Kreuzinger N (2004) Comparison of the behaviour of selected micropollutants in a membrane bioreactor and a conventional wastewater treatment plant. Water Sci and Technol 50(5):29–36
8. Cirja M, Ivashechkin P, Schaffer A, Corvini PFX (2008) Factors affecting the removal of organic micropollutants from wastewater in conventional treatment plants (CTP) and membrane bioreactors (MBR). Rev Environ Sci Biotechnol 7(1):61–78
9. Bernhard M, Muller J, Knepper TP (2006) Biodegradation of persistent polar pollutants in wastewater: comparison of an optimised lab-scale membrane bioreactor and activated sludge treatment. Water Res 40:3419–3428
10. Kimura K, Hara H, Watanabe Y (2005) Removal of pharmaceutical compounds by submerged membrane bioreactors (MBRs). Desalination 178:135–140
11. Radjenovic J, Petrovic M, Barceló D (2007) Analysis of pharmaceuticals in wastewater and removal using a membrane bioreactor. Anal Bioanal Chem 387:1365–1377
12. Radjenovic J, Petrovic M, Barceló D (2009) Fate and distribution of pharmaceuticals in wastewater and sewage sludge of the conventional activated sludge (CAS) and advanced membrane bioreactor (MBR) treatment. Water Res 43(3):831–841
13. Lishman L, Smyth SA, Sarafin K, Kleywegt S, Toito J, Peart T (2006) Occurrence and reduction of pharmaceuticals and personal care products and estrogens by municipal wastewater treatment plants in Ontario, Canada. Sci Tot Environ 367:544–558
14. Vieno N, Tuhkanen T, Kronberg N (2007) Elimination of pharmaceuticals in sewage treatment plants in Finland. Water Res 41:1001–1012
15. Zhang Y, Geissen SU, Gal C (2008) Carbamazepine and diclofenac: removal in wastewater treatment plants and occurrence in water bodies. Chemosphere 73:1151–1161

16. Joss S, Zabczynski A, Gobel B, Hoffmann D, Loffler CS, McArdell TA (2006) Biological degradation of pharmaceuticals in municipal wastewater treatment: proposing a classification scheme. Water Res 4:1686–1696
17. Batt AL, Kim S, Aga DS (2006) Enhanced biodegradation of iopromide and trimethoprim in nitrifying activated sludge. Environt Sci Tehnol 40:7367–7373
18. Perez S, Eichhorn P, Aga DS (2005) Evaluating the biodegradability of sulphamenthazine, sulphamethoxzole and trimethoprin at different stages of sewage treatment. Environ Toxicol Chem 24(6):1361–1367
19. Auriol M, Filali-Meknassi Y, Tyagi RD, Adams CD, Surampalli RY (2006) Endocrine disrupting compounds removal from wastewater, a new challenge. Process Biochem 41(3):525–539
20. Koh YKK, Chiu TY, Boobis A, Cartmell E, Scrimshaw MD, Lester JN (2008) Treatment and removal strategies for estrogens from wastewater. Environ Technol 29(3):245–267
21. Bjorkblom C, Salste L, Katsiadaki I, Wiklund T, Kronberg L (2008) Detection of estrogenic activity in municipal wastewater effluent using primary cell cultures from three-spined stickleback and chemical analysis. Chemosphere 73(7):1064–1070
22. Hjelmborg PS, Ghisari M, Bonefeld-Jorgensen EC (2006) SPE-HPLC purification of endocrine-disrupting compounds from human serum for assessment of xenoestrogenic activity. Anal Bioanal Chem 385(5):875–887
23. Mnif W, Dagnino S, Escande A, Pillon A, Fenet H, Gomez E, Casellas C, Duchesne M, Hernandez-Raquet G, Cavaillés V, Balaguer P, Bartegi A (2010) Biological analysis of endocrine-disrupting compounds in Tunisian sewage treatment plants. Arch Environ Contam Toxicol 59(1):1–12
24. Fernandez MP, Buchanan ID, Ikonomou MG (2008) Seasonal variability of the reduction in estrogenic activity at a municipal WWTP. Water Res 42(12):3075–3081
25. Hashimoto T, Onda K, Nakamura Y, Tada K, Miya A, Murakami T (2007) Comparison of natural estrogen removal efficiency in the conventional activated sludge process and the oxidation ditch process. Water Res 41(10):2117–2126
26. Mispagel C, Allinson G, Allinson M, Shiraishi F, Nishikawa M, Moore MR (2009) Observations on the estrogenic activity and concentration of 17β-estradiol in the discharges of 12 wastewater treatment plants in southern Australia. Arch Environ Contam Toxicol 56(4):631–637
27. Svenson A, Allard AS (2003) Removal of estrogenicity in Swedish municipal sewage treatment plants. Water Res 37(18):4433–4443
28. Céspedes R, Petrovic M, Raldúa D, Saura U, Piña B, Lacorte S, Viana P, Barceló D (2003) Integrated procedure for determination of endocrine-disrupting activity in surface waters and sediments by use of the biological technique recombinant yeast assay and chemical analysis by LC–ESI-MS. Anal Bioanal Chem 378(3):697–708
29. Creusot N, Kinani S, Balaguer P, Tapie N, LeMenach K, Maillot-Marechal E, Porcher JM, Budzinski H, Ait-Aissa S (2010) Evaluation of an hPXR reporter gene assay for the detection of aquatic emerging pollutants: screening of chemicals and application to water samples. Anal Bioanal Chem 396(2):569–583
30. Fernandez MP, Noguerol TN, Lacorte S, Buchanan I, Pina B (2009) Toxicity identification fractionation of environmental estrogens in waste water and sludge using gas and liquid chromatography coupled to mass spectrometry and recombinant yeast assay. Anal Bioanal Chem 393(3):957–968
31. Harris CA, Henttu P, Parker MG, Sumpter JP (1997) The estrogenic activity of phthalate esters in vitro. Environ Health Perspect 105(8):802–811
32. Isobe T, Shiraishi H, Yasuda M, Shinoda A, Suzuki H, Morita M (2003) Determination of estrogens and their conjugates in water using solid-phase extraction followed by liquid chromatography–tandem mass spectrometry. J Chromatography A 984(2):195–202
33. Jugan ML, Oziol L, Bimbot M, Huteau V, Tamisier-Karolak S, Blondeau JP, Levi Y (2009) In vitro assessment of thyroid and estrogenic endocrine disruptors in wastewater treatment

plants, rivers and drinking water supplies in the greater Paris area (France). Sci Tot Environ 407(11):3579–3587
34. Korner W, Vinggaard AM, Terouanne B, Ma R, Wieloch C, Schlumpf M, Sultan C, Soto AM (2004) Interlaboratory comparison of four in vitro assays for assessing androgenic and antiandrogenic activity of environmental chemicals. Environ Health Perspect 112(6):695–702
35. Sousa A, Schonenberger R, Jonkers N, Suter MJF, Tanabe S, Barroso CM (2010) Chemical and biological characterisation of estrogenicity in effluents from WWTPs in Ria de Aveiro (NW Portugal). Arch Environ Contam Toxicol 58(1):1–8
36. Tan BLL, Hawker DW, Muller JF, Leusch FDL, Tremblay LA, Chapman HF (2007) Comprehensive study of endocrine disrupting compounds using grab and passive sampling at selected wastewater treatment plants in South East Queensland, Australia. Environ Int 33(5):654–669
37. Bertanza G, Pedrazzani R, Dal Grande M, Papa M, Zambarda V, Montani C, Steimberg N, Mazzoleni G, Di Lorenzo D (2011) Effect of biological and chemical oxidation on the removal of estrogenic compounds (NP and BPA) from wastewater: an integrated assessment procedure. Water Res 45:2473–2484
38. Clara M, Kreuzinger N, Strenn B, Gans O, Kroiss H (2005) The solids retention time—a suitable design parameter to evaluate the capacity of wastewater treatment plants to remove micropollutants. Water Res 39(1):97–106
39. Koh YKK, Chiu TY, Boobis A, Scrimshaw MD, Bagnall JP, Soares A, Pollard S, Cartmell E, Lester JN (2009) Influence of operating parameters on the biodegradation of steroid estrogens and nonylphenolic compounds during biological wastewater treatment processes. Environm Sci Technol 43(17):6646–6654
40. Cicek N, Franco JP, Suidan MT, Urbain V, Manem J (1999) Characterisation and comparison of a membrane bioreactor and a conventional activated-sludge system in the treatment of wastewater containing high-molecular-weight compounds. Water Environ Res 71(1):64–70
41. Clouzot L, Doumenq P, Roche N, Marrot B (2010) Kinetic parameters for 17 alpha-ethinylestradiol removal by nitrifying activated sludge developed in a membrane bioreactor. Bioresour Technol 101(16):6425–6431

Chapter 4
Application of Wet Oxidation to Remove Trace Pollutants from Wastewater

Verónica García-Molina and Santiago Esplugas

Abstract This chapter evaluates the application of Wet Oxidation (WO) for the treatment of aqueous effluents to remove pollutants. WO is a destructive wastewater technology based on the oxidation of pollutants at moderate high temperature (130 ± 300 °C) and pressures (5 ± 200 bar) in the liquid phase. The organic material is not normally completely destroyed, but converted to intermediate end products with a significant reduction in total organic carbon and chemical oxygen demand. The final aqueous effluent will contain a considerable quantity of low molecular weight organics, ammonia, inorganic acids and inorganic salts being more biodegradable than the untreated effluent. The economy of the water treatment process is thus highly optimized by combining WO with a conventional biological treatment.

Keywords Wet oxidation · Wastewater · Advanced oxidation process · Pollutant

4.1 Introduction

The need to restore wastewater to avoid further damage to the environment has arisen in the last years the development of effective methods for pollutants removal. The main goal is to attain complete mineralization to CO_2 and H_2O in

V. García-Molina
Dow Water & Process Solutions, Dow Chem Iberica SL,
Tarragona, Spain
e-mail: vgarciamolina@dow.com

S. Esplugas (✉)
Department of Chemical Engineering, University of Barcelona,
Martí i Franquès 1, 08028, Barcelona, Spain
e-mail: santi.esplugas@ub.edu

G. Lofrano (ed.), *Green Technologies for Wastewater Treatment*,
SpringerBriefs in Green Chemistry for Sustainability,
DOI: 10.1007/978-94-007-1430-4_4, © García-Molina, Esplugas 2012

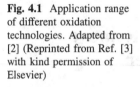

Fig. 4.1 Application range of different oxidation technologies. Adapted from [2] (Reprinted from Ref. [3] with kind permission of Elsevier)

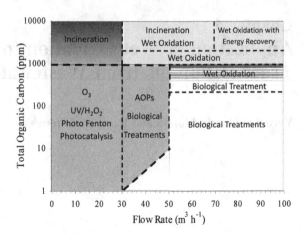

addition to small amounts of some ions, e.g. chloride anions, or at least to produce less harmful intermediates. An ideal waste treatment process must completely mineralize the toxic species present in the waste streams without leaving behind any hazardous residues and should be cost-effective as well. The so-called Advanced Oxidation Processes (AOPs) appear to be a promising field of study as wastewater treatments, for the reason that the organic components that are thermodynamically unstable to the oxidation are eliminated and not transferred from one phase to another [1].

There are several technologies classified as AOPs and each one is at a different level of development and commercialization. Figure 4.1 shows the application range of some wastewater treatments depending on the flow rate and organic matter content of the effluent to be treated. According to this illustration, technologies based on UV radiation and ozonation should be preferred at low flow rates and low organic loads in the incoming effluent. When the incoming effluent contains a high organic load, processes such as incineration and wet oxidation should be employed depending on the flow rate of the effluent. On the other hand, biological treatments appear to be suitable when the flow rate of the feed is high and it has a low content of organic matter.

The first wet oxidation patent was obtained in 1950 by Zimmermann [4], although it was already discovered in 1935. It was first used as a completely new method of obtaining vanillin directly from pulping liquor by partial oxidation of the ligno-sulfonic acids. The technology was introduced to the pulp and paper market in 1955 and to the municipal sewage sludge market in the late-1950s and early-1960s. The process can treat any kind of waste, produced by various branches of industrial activity or sludge produced by conventional treatment processes [5]. Nowadays, the main uses of this technology are:

1. Treatment of spent caustic liquors that typically come from plants of ethylene production (from the scrubbing of cracked gas with aqueous sodium hydroxide) or from oil refining plants (from the extraction or treatment of acidic impurities,

such as hydrogen sulfide, mercaptans and organic acids in hydrocarbon streams).

2. Treatment of high strength waste streams in order to make them more suitable to conventional treatments such as biological polishing, or as pretreatment for product recovery. Wet oxidation destroys the large molecules in waste, converting them predominantly to carbon dioxide with some formation of low weight carboxylic acids such as acetic acid, which is highly biodegradable. The purpose of this treatment is to condition a waste streams that is: toxic, reactive, refractory to biotreatment or hazardous.

3. Treatment of sludge that includes:

   • Sludge dewatering: Low pressure/temperature oxidation is used for sludge conditioning to allow its dewatering.
   • Sludge destruction: At higher temperatures, volatiles in sludge can be destroyed.
   • Wet Air Regeneration: Wet air oxidation is used in conjunction with a biological process referred to as the powdered activated carbon treatment system for both regeneration of carbon and destruction of biological sludge.

According to manufactures information, currently there are more than 200 wet oxidation plants in the world that are operated for the treatment of different types of waste streams, such as effluents from pulp and paper mills, wastewater from petrochemical plants, textile wastewaters, thermo-mechanical pulp sludge, paper mill sludge, sewage sludge, activated sludge, blow down effluents from crystallizers and so forth. Wet air oxidation also finds its application in the petrochemical and textile industries, where ultrasonic irradiation in the presence of Cu^{2+} ions is employed as an effective pretreatment of the process [6].

4.2 The Wet Oxidation Process as Green Technology

Wet oxidation involves the liquid phase oxidation of organic or oxidizable inorganic components at elevated temperatures and pressures using a gaseous source of oxygen. This technology is commonly named wet oxidation (WO) when pure oxygen is used, wet air oxidation (WAO) when air is supplied to the system and wet peroxidation (WPO) when hydrogen peroxide is used as oxidant instead oxygen. Additionally carbon materials, noble metals such as Ru, Rh, Pd, Ir, and Pt as well as oxides of Cr, Mn, Fe, Co, Ni, Cu, Zn, Mo, and Ce have been used heterogeneous catalysts to enhanced the degradation [7, 8]. In the scientific literature it is possible to find interesting reviews on catalytic and noncatalytic wet oxidation [9–11]. Phenolics, carboxylic acids and nitrogen-containing compounds have been taken as model pollutants in WO treatments [12–15] and some scientific information of their application to wastewaters may be found [3, 16–26]. Pulp and paper mill effluents, agro-food streams, dyeing wastewater concentrates and leachates from solid waste have been treated by WO.

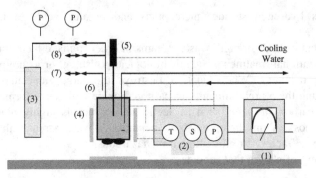

Fig. 4.2 Wet Oxidation lab equipment: *1* power supply, *2* digital controller (Temperature, Pressure and Stirring Speed), *3* oxygen bottle, *4* heater, *5* stirrer, *6* reactor, *7* samples extraction, *8* gas draining

As it is possible to appreciate in Fig. 4.2, the equipment at laboratory scale for Wet Oxidation experimentation is very simple. However it is very important to have a good control of temperature, pressure and stirring speed in the oxidation chamber. High-pressure stainless steel reactor with a volume around 500 mL is normally used. This simple wet oxidation reactor permits to conduct experiments at neutral or close to acidic pH and is capable of performing batch experiments at pressures of up to 50 bar and temperatures of up to 623 K. A drop band with one screw and a split ring pair with screws allow the use of the reactor under those conditions of pressure and temperature.

The WO processes have a very limited interaction with the environment and when the oxidation is not complete it can be coupled with a biological treatment to eliminate or to treat any kind of waste, even toxic [15, 19, 24, 27].

The following schematic reactions represent the fundamental transformations of organic matter over the duration of the wet oxidation process [28, 29]:

$$Organics + O_2 \rightarrow CO_2 + H_2O + RCOOH \tag{4.1}$$

$$Sulphur\ Species + O_2 \rightarrow SO_4^{-2} \tag{4.2}$$

$$Organic\ Cl \rightarrow Cl^- + CO_2 + RCOOH \tag{4.3}$$

$$Phosphorous + O_2 \rightarrow PO_4^{-3} \tag{4.4}$$

$$Organic\ N + O_2 \rightarrow CO_2 + H_2O + NH_3 \left(or/and\ N_2,\ NO_3^-\right) \tag{4.5}$$

From the previous scheme it can be noted that over the duration of the WO, the organic compounds are reduced to CO_2 or other innocuous components, nitrogen is transformed into NH_3, NO_3 or elementary nitrogen and finally, halogen compounds and sulfurs are transformed into halides and sulfates [30]. It should be mentioned that WO presents an important advantage compared with other processes, since no NO_x, SO_2, HCl, dioxides or other harmful products are generated within the process.

Fig. 4.3 4CP concentration and TOC changes during WO treatment of a 500 ppm aqueous solution versus partial pressure of oxygen [18]

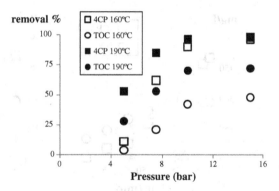

Besides the pollutant concentration, there are two parameters commonly used to characterize the efficiency of the wet oxidation process are the Total Organic Carbon (TOC) and the Chemical Oxygen Demand (COD). The comparison between TOC or COD, measured at the commencement and at the end of the reaction, allows the knowledge of the degree of mineralization of the process or in other words, the amount of organic matter transformed into CO_2, which is normally one of the main objectives to be accomplished. A typical COD removal achieved after a wet oxidation oscillates between 75 and 90% [31]. Operating conditions in WO are very important as could be observed in Fig. 4.3, where removal percentage of 4-CP (chlorophenol) and TOC at the end of a WO treatment are presented at different temperature and partial pressure of oxygen. The increase of temperature and partial pressure of oxygen results in a higher the 4-CP and TOC removal percentage. However, despite attaining a 100% of 4-CP removal, only a 70% of TOC removal is reached, indicating thus, the presence of intermediate compounds (most likely low molecular weight acids) at the end of the process. As it can be observed in Fig. 4.4 the operating temperature has a strong influence on the removal.

The oxidation of the initial matter is not always complete and the result of the reaction is a mixture of biodegradable products of low molecular weight such as organic acids, aldehydes and alcohols. The non-complete oxidation can be explained taking into account that the oxidation rate increases along with the increase in the molecular weight/carbon number [29]. As a consequence, low molecular weight acids, which are the last organic intermediates formed throughout the process previous to the formation of carbon dioxide, are the most refractory compounds for the oxidation process and remain in the solution. The formation of these carboxylic acids causes on one side the decrease of the pH over the duration of the process and on the other hand an increase in the biodegradability of the initial water stream. In Fig. 4.5 the evolution of some parameters such as COD, Biochemical Oxygen Demand (BOD) of a wastewater effluent from a Finnish Pulp and Paper mill [18] is depicted versus the treatment time. It can be observed that the BOD of the solution first increases, then reaches a maximum and finally decreases.

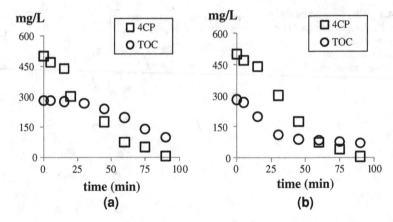

Fig. 4.4 4CP concentration and TOC changes during WO treatment of a 500 ppm aqueous solution at 10 bar. **a** 175 °C, **b** 190 °C [18]

Fig. 4.5 BOD–COD changes during WO treatment of a wastewater effluent from a Finnish Pulp and Paper mill [18]

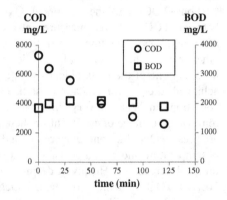

Fig. 4.6 BOD–COD changes during WO treatment of a 500 ppm 4-CP at 10 bar and 190 °C [18]

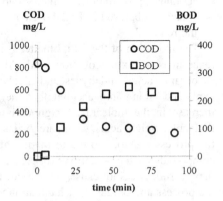

    The initial increase is due to the degradation of the non-biodegradable organic compounds and the formation of biodegradable intermediates. This period of time is characterized by having an increasing ratio: biodegradable

Table 4.1 Operating conditions in WO systems

Reference	Temperature (°C)	Pressure (bar)[a]
Li et al. [31]	150–350	20–200
Mishira et al. [29]	125–320	5–200
Beyrich et al. [32]	150–300	50–200
Escalas et al. [33]	175–320	22–208
Debellefontaine and Foussard [27]	200–325	Up to 150
Perkow et al. [30]	150–330	30–250
Foussard et al. [34]	197–327	20–200

[a] The pressure in the reactor is the sum of the pressure of the steam generated at this temperature and the pressure of the air or oxygen supplied to maintain an elevated oxygen concentration in the liquid phase

matter/non-biodegradable matter. At some point, the non-biodegradable matter present in the solution reaches its minimum because it has been widely oxidized. This moment coincides with the maximum biodegradability. After reaching this maximum, the oxidation of the biodegradable intermediate compounds takes place and since the amount of biodegradable matter decreases, the biodegradability of the solution also decreases until the end of the reaction. When an aqueous solution of a pollutant is treated by WO the behavior is similar as it is shown in Fig. 4.6 where a 500 ppm 4-CP aqueous solution is treated by WO at 10 bar and 190 °C.

The degree of oxidation depends, on one side on the operating conditions i.e. temperature, pressure and residence time and on the other side, on the organic compounds resistance to chemical oxidation. The range of temperature and pressure at which the reaction is carried out is not strictly limited. However, the pressure should be always maintained well above the saturation pressure corresponding to the operating temperature, so that the reaction occurs in the liquid phase. Table 4.1 shows some of the conditions of pressure and temperature found in the literature. The usual times of reaction are between 15 and 120 min [31].

When evaluating the effluent to be treated it has to be noted that wet oxidation is a reaction accompanied by a release of energy [5] and, thus, in order for the process to be energy self-sufficient, the COD of the wastewater should be high. Several authors have reported an optimum value between 10 and 20 kg/m^3 of COD in the entry stream.

Compared with other treatments such as incineration, wet oxidation requires much less energy. The fact is that for WO the only energy required is the difference in enthalpy between the incoming and outcoming, whereas, for incineration, not only the sensible enthalpy is to be provided but also the heat for the complete evaporation of water [29]. The autogenous point for wet oxidation corresponds to a stream inlet of 10 g/L of organic matter [35], whereas a feed stream of more than 200 g/L is necessary in order to reach autogenous conditions in incineration. At this point, where incineration becomes autogenous, wet oxidation is already extensively exothermic.

On the other hand, the capital costs of a WO system are high and depend on the flow and oxygen demand of the effluent, severity of the oxidation conditions, and the required construction materials. The reactor itself can account for a significant fraction (50%) of the total equipment cost. [29].

4.3 Kinetics of Wet Oxidations

For a specific pollutant a first order reaction with respect to the pollutant concentration normally provides a consistent model. However, some pollutants exhibit an induction period, the length of which depends on the oxygen partial pressure, followed by a fast reaction step. Thus, the reaction can be divided in two separate parts. In the first one, i.e. induction period, the radicals are formed and in the second part, the oxidation takes place.

For a complex and multi-component wastewaters, the prediction of the evolution of certain key parameters such as TOC, COD or TOD is not as simple as with waters containing a unique pollutant. As previously stated, as a result of the variation in the original components or pollutants concentrations during the WO process, important changes in lumped parameters as TOC, COD and Total Oxygen Demand (TOD) occur. Many kinetics models for multi-compound solutions have been suggested in the literature. One of first ones was the General Lumped Kinetic Model (GLKM) suggested by Li et al. [31]. After some predictions and considerations, the mathematical model leads to Eq. 4.6, which allows the prediction of the COD, TOC or TOD over the duration of the reaction:

$$\frac{[A + B]}{[A + B]_o} = \frac{[A]_o}{[A]_o + [B]_o} \left[\frac{k_2}{k_1 + k_2 - k_3} e^{-k_3 t} + \frac{(k_1 - k_3)}{k_1 + k_2 - k_3} e^{-(k_1 + k_2)t} \right]$$
$$+ \frac{[B]_o}{[A]_o + [B]_o} e^{-k_3 t}$$

$$(4.6)$$

where A represents all initial and relatively unstable intermediate organic compounds except acetic acid and B contains the refractory intermediates represented by acetic acid. The model assumes a scheme of the reaction pathways as shown in Fig. 4.7.

In addition, the model assumes the following considerations:

(1) The concentration of the groups A or B may be expressed in forms of TOC, COD or TOD.
(2) Based on the bibliography, the reaction rate may be assumed to be first order to group A or B, and nth order to oxygen.
(3) The reactor is supposed to follow the model of an isothermal and ideal batch reactor or plug-flow reactor with constant volumetric flow rate.

Some other models have been found in the literature, such as the Lumped Kinetic Model by Zhang and Chuang [36], the Multi-component Kinetic Model

Fig. 4.7 Scheme of the reaction pathways

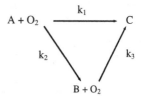

Fig. 4.8 Scheme of the WO reaction mechanism suggested by the model of Verenich and Kallas [24]

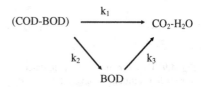

suggested by Escalas et al. [33], the Lumped Kinetic Model for Oil Waters by López Bernal et al. [37], the Extended Kinetic Model by Belkacemi et al. [16], and the Wet Oxidation Lumped Kinetic Model for Wastewater Organic Burden Biodegradability Prediction by Verenich and Kallas [24].

The model proposed by Verenich and Kallas [24] evaluates the changes in the biodegradability of the wastewater during the oxidation. These changes are important for the applicability of further biological processing, which is an economic way to reduce the organic load of the water. The model, based on these qualities, assumes the WO mechanism depicted in Fig. 4.8.

According to this reaction pathway, the difference between the total organics measured via COD and biodegradable compounds described by BOD represents the refractory initial organic compounds. Their oxidation is assumed to proceed in two parallel ways: through the first one, they are oxidized to the end products of the reaction; through the second one, the matter contained in the waste stream is partially oxidized and turns into biodegradable. The biodegradable compounds are further oxidized to end products, i.e. carbon dioxide and water. When comparing this model with the one proposed by Li et al. (previously explained) it is important to notice that both kinetics models are based on a three-reaction scheme, however they differ in the description of the compounds contained in each one of the reactant groups and consequently, the equations describing both models are similar but not equal. One common characteristic is that both models assume that the reactor follows the model of an isothermal and ideal batch reactor. Verenich and Kallas [24] suggested that the oxidation rate of each category of organics follows a pseudo-first order kinetic rate, which is in agreement with the models presented by Li et al. [31] and Zhang and Chuang [36]. Equations 4.7 and 4.8 show the obtained variation of COD and BOD during the reaction time using three first kinetics constants.

$$[COD] = [BOD] + [(COD - BOD)]_o e^{-(k_1 + k_2)t} \qquad (4.7)$$

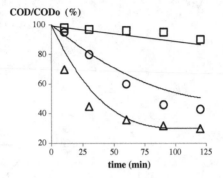

Fig. 4.9 Experimental and calculated values of the ratio COD/COD$_0$. WO treatment of a wastewater effluent from a Finnish Pulp and Paper mill at the pO_2 bar. Values obtained experimentally (*triangles* 160 °C, *circles* 180 °C, *squares* 200 °C) and theoretically (*continuous lines*) using simulation software Mathematica and kinetic model of Li et al. [31]

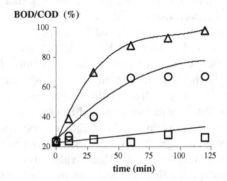

Fig. 4.10 Experimental and calculated values of the ratio BOD/COD. WO treatment of a wastewater effluent from a Finnish Pulp and Paper mill at the pO_2 bar. Values obtained experimentally (*triangles* 160 °C, *circles* 180 °C, *squares* 200 °C) and theoretically (*continuous lines*) using simulation software Mathematica and kinetic model of Verenich and Kallas [24]

$$[BOD] = [BOD]_o e^{-k_3 t} + \frac{k_2[(COD - BOD)]_o}{k_1 + k_2 - k_3}\left(e^{-k_3 t} - e^{-(k_1+k_2)t}\right) \quad (4.8)$$

Figures 4.9 and 4.10 shows respectively the experimental and calculated values of the ratio COD/COD$_0$ and the ratio BOD/COD for a WO treatment of a wastewater effluent from a Finnish Pulp and Paper mill at the pO_2 bar at different temperatures and using the kinetics model of Li et al. [31] and Verenich and Kallas [24]. It can be observed that the importance of the temperature in the removal of COD and in the increasing of the biodegradability measured as the ratio BOD/COD. In both figures it can be seen the good agreement between the mathematical modeling and the experimental results obtained.

4.4 Concluding Remarks

Wet oxidation is a promising technology to treat liquid wastes with moderate to high organic content (i.e. 10–100 g/L COD). Even though it is a relatively new technology, it is already in full scale operation in more than 200 installations worldwide. Among the various characteristics of this technique, the more representative are its high efficiency, its flexibility and the possibility of customizing the process with the use of different catalysts in order to obtain the highest efficiency rates depending on the original stream to be treated. An additional key advantage of wet oxidation is that it is a destructive technology, i.e., the pollutants are eliminated and not transferred from one phase to another.

Wet Oxidation can usually reach complete mineralization of the original stream by optimizing the various operating conditions (time, temperature and oxygen partial pressure) but, in order to improve the overall economy of the process the reaction should be stopped at the point where the maximum biodegradability is reached. The treatment is then completed with the addition of a post-biological treatment, which has lower operating costs than Wet Oxidation.

Acknowledgments   Authors are grateful to Spanish Ministry of Education and Science (CICYT Project CTQ2008-01710 and Consolider-Ingenio NOVEDAR 2010 CSD2007-00055) for funds received to carry out this work.

References

1. Glaze WH, Kang JW, Chapin DH (1987) The chemistry of water treatment processes involving ozone, hydrogen peroxide and UV-radiation. Ozone Sci Eng 9:335
2. Hancock FE (1999) Catalytic strategies for industrial water reuse. Catal Today 53:3
3. Gotvajn AZ, Zagorc-Koncan J, Tisler T (2007) Pretreatment of highly polluted pharmaceutical waste broth by wet air oxidation. J Environ Eng ASCE 133:89
4. Zimpro (2011) Wet air oxidation (WAO) and wet oxidation systems. http://www.water.siemens.com/en/products/physical_chemical_treatment/hydrothermal_oxidation_wao/Pages/Zimpro_Wet_Air_Oxidation.aspx. Accesssed 14 March 2011
5. Debellefontaine H, Chakchouk M, Foussard JN, Tissot D, Striolo P (1996) Treatment of organic aqueous wastes: wet air oxidation and wet peroxide oxidation. Environ Pollut 92:155
6. Papadaki M, Emery RJ, Abu-Hassan MA, Díaz-Bustos A, Metcalfe IS, Mantzavinos D (2004) Sonocatalytic oxidation processes for the removal of contaminants containing aromatic rings from aqueous effluents. Sep Purif Technol 34:35. doi:10.1016/S1383-5866(03)00172-2
7. Kim K, Ihm S (2011) Heterogeneous catalytic wet air oxidation of refractory organic pollutants in industrial wastewaters: a review. J Hazard Mater 186:16. doi:10.1016/j.jhazmat.2010.11.01
8. Stuber F, Font J, Fortuny A, Bengo C, Eftaxias A, Fabregat A (2005) Carbon materials and catalytic wet air oxidation of organic pollutants in wastewater. Top Catal 33:3. doi:10.1007/s11244-005-2497-1
9. Bhargava SK, Tardio J, Prasad J, Föger K, Akolekar DB, Grocott SC (2006) Wet oxidation and catalytic wet oxidation. Ind Eng Chem Res 45:1221. doi:10.1021/ie051059n

10. Levec J, Pintar A (2007) Catalytic wet-air oxidation processes: a review. Catalysis Today 124:172. doi:10.1016/j.cattod.2007.03.03
11. Luck F (1999) Wet air oxidation: past, present and future. Catal Today 53:81
12. García-Molina V, Lopez-Arias M, Florczyk M, Chamarro E, Esplugas S (2005) Wet peroxide oxidation of chlorophenols. Water Res 39:795. doi:10.1016/j.watres.2004.12.00
13. Lopes RJG, Silva AMT, Quinta-Ferreira RM (2007) Screening of catalysts and effect of temperature for kinetic degradation studies of aromatic compounds during wet oxidation. Appl Catal B Environ 73:193. doi:10.1016/j.apcatb.2006.11.01
14. Santos A, Yustos P, Cordero T, Gomis S, Rodriguez S, García-Ochoa F (2005) Catalytic wet oxidation of phenol on active carbon: stability, phenol conversion and mineralization. Catal Today 102–103:213. doi:10.1016/j.cattod.2005.02.00
15. Suarez-Ojeda ME, Fabregat A, Stuber F, Fortuny A, Carrera J, Font J (2007) Catalytic wet air oxidation of substituted phenols: temperature and pressure effect on the pollutant removal, the catalyst preservation and the biodegradability enhancement. Chem Eng J 132:105. doi:10.1016/j.cej.2007.01.02
16. Belkacemi K, Larachi F, Hamoudi S, Turcotte G, Sayari A (1999) Inhibition and deactivation effects in catalytic wet oxidation of high-strength alcohol-distillery liquors. Ind Eng Chem Res 38:2268. doi:10.1021/ie980005t
17. Chen G, Lei L, Hu X, Yue PL (2003) Kinetic study into the wet air oxidation of printing and dyeing wastewater. Sep Purif Technol 31:71
18. Garcia-Molina V (2006) Wet oxidation processes for water pollution remediation. PhD thesis. department of chemical engineering. faculty of chemistry. University of Barcelona. http://www.tdx.cesca.es/TESIS_UB/AVAILABLE/TDX-0804106-10239/VGM_Doc-toral_Thesis.pdf . Accessed 14 Mar 2011
19. Garg A, Mishra A (2010) Wet oxidation: an option for enhancing biodegradability of leachate derived from municipal solid waste (MSW) landfill. Ind Eng Chem Res 49:5575. doi:10.1021/ie100003q
20. Lei L, Hu X, Chen G, Porter JF, Yue PL (2000) Wet air oxidation of desizing wastewater from the textile industry. Ind Eng Chem Res 39:2896. doi:10.1021/ie990607s
21. Lei L, Hu X, Yue PL (1998) Improved wet oxidation for the treatment of dyeing wastewater concentrate from membrane separation process. Water Res 32:2753
22. Lin SH, Ho SJ, Wu CL (1996) Kinetic and performance characteristics of wet air oxidation of high-concentration wastewater. Ind Eng Chem Res 35:307. doi:10.1021/ie950251u
23. Perathoner S, Centi G (2005) Wet hydrogen peroxide catalytic oxidation (WHPCO) of organic waste in agro-food and industrial streams. Top Catal 33:207. doi:10.1007/s11244-005-2529-x
24. Verenich S, Kallas J (2002) Biodegradability enhancement by wet oxidation in alkaline media: delignification as a case study. Environ Technol 23:655
25. Verenich S, Roosalu K, Hautaniemi M, Laari A, Kallas J (2005) Kinetic modeling of the promoted and unpromoted wet oxidation of debarking evaporation concentrates. Chem Eng J 108:101. doi:10.1016/j.cej.2005.01.00
26. Zhu W, Bin Y, Li Z, Jiang Z, Yin T (2002) Application of catalytic wet air oxidation for the treatment of H-acid manufacturing process wastewater. Water Res 36:1947
27. Debellefontaine H, Foussard JN (2000) Wet air oxidation for the treatment of industrial wastes. Chemical aspects, reactor design and industrial applications in Europe. Waste Manag 20:15
28. Kolaczkowski ST, Plucisnki P, Beltrán F, Rivas FJ, McLurg DB (1999) Wet air oxidation: a review of process technologies and aspects in reactor design. Chem Eng J 73:143
29. Mishira VS, Mahajani VV, Joshi JB (1995) Wet air oxidation. Env Eng Chem Res 34:2
30. Perkow H, Steiner R (1981) Wet air oxidation: a review. Ger Chem Eng 4:193
31. Li L, Peishi C, Earnest FG (1991) Generalized kinetic model for wet oxidation of organic compounds. AIChE J 37:1687
32. Beyrich J, Gautschi W, Regenass W, Wiedmann W (1979) Design of reactors for the wet air oxidation of industrial wastewater by means of computer simulation. Comp Chem Eng 3:161

33. Escalas A, González M, Baldasano JM, Gassó S (1997) A multicomponent kinetic model for wet oxidation. Chem Oxid Technol Nineties 5:39
34. Foussard J, Debellefontaine H, Besombes-Vailhé J (1989) Efficient elimination of organic liquid wastes: wet air oxidation. Env Eng 115:367
35. Wilhelmi AR, Knopp PV (1979) Wet air oxidation—an alternative to incineration. Chem Eng Prog 46:46
36. Zhang Q, Chuang KT (1996) Lumped kinetic model for catalytic wet oxidation of organic compounds in industrial wastewater. AIChE J 45:145
37. Lopez-Bernal J, Portela Miguélez JR, Nebot Sanz E, Martínez de la Ossa E (1999) Wet air oxidation of oily wastes generated aboard ships: kinetic modeling. J Hazard Mater B 67:61

Chapter 5
Advanced Oxidation of Endocrine Disrupting Compounds: Review on Photo-Fenton Treatment of Alkylphenols and Bisphenol A

Idil Arslan-Alaton and Tugba Olmez-Hanci

Abstract The discharge of man-made chemicals into aquatic environment creates an ever-increasing challenge to scientists and engineers. These chemicals can harm living organisms even at below ppb levels and hence only the most effective treatment processes should be employed for their destruction and detoxification. The application of Photo-Fenton process and complimentary treatment systems (H_2O_2/UV-C and Fenton's reagent) for the degradation of two industrial pollutant categories with significant endocrine disrupting properties, namely the alkyl phenols nonyl and octyl phenol as well as bisphenol A, has been discussed and reviewed in the present chapter.

Keywords Endocrine disrupting compounds (EDCs) · Bisphenol A (BPA) · Nonylphenol (NP) · Octylphenol (OP) · Photo-Fenton process

5.1 Introduction

Advanced Oxidation Processes (AOPs) are being employed to treat biologically inert, hazardous, toxic and other problematic pollutants found in air, water and wastewater. Among the variety of AOPs available, Photo-Fenton (Fe^{2+}/H_2O_2/UV) treatment systems have gained more attention than the others due to their superior

I. Arslan-Alaton (✉) · T. Olmez-Hanci
Civil Engineering Faculty, Environmental Engineering Department,
Istanbul Technical University, Ayazaga Kampusu, 34469, Maslak,
Istanbul, Turkey
e-mail: arslanid@itu.edu.tr

T. Olmez-Hanci
e-mail: tolmez@itu.edu.tr

G. Lofrano (ed.), *Green Technologies for Wastewater Treatment*,
SpringerBriefs in Green Chemistry for Sustainability,
DOI: 10.1007/978-94-007-1430-4_5, © Arslan-Alaton, Olmez-Hanci 2012

reaction rates and efficiencies, technical feasibility and attractive processes economics. The principles and reaction mechanisms of iron-based photochemical treatment systems have been extensively studied in the past starting from the late 1980s to early 1990s. Extensive research on reaction pathways and critical process parameters were followed by the development of heterogeneous Fenton systems. More recently, iron-based AOPs have been applied to degrade priority pollutants that create a potential risk in the environment due to their hormone mimicking and hence endocrine disrupting effects. These chemicals have negative effects on living organisms even at below ppb levels and hence only the most effective AOPs can be employed for their effective destruction and detoxification. The present chapter reviews the application of Photo-Fenton process and complimentary treatment systems (H_2O_2/UV-C and Fenton's reagent) for the degradation of two industrial pollutants with significant endocrine disrupting properties; alkyl phenols (nonyl and octyl phenols) and bisphenol A.

5.2 The Photo-Fenton Treatment Process

5.2.1 The Photo-Fenton Reaction: General Overview

In the past two decades, photochemical processes for the destruction of refractory and/or toxic organic pollutants found in water or wastewater have gained significant interest. The hydroxyl radical ($HO^\bullet$) is being accepted as the major free radical that reacts and degrades aqueous pollutants almost indiscriminately. There are several methods employed to produce $HO^\bullet$, among which H_2O_2/UV-C and O_3/UV-C are the most well-known and established ones. Another homogenous, photochemical advanced oxidation process (AOP) is the Photo-Fenton reaction, where $HO^\bullet$s are initially produced in three ways; (i) photoreduction of ferric iron hydroxo complexes to ferrous iron, (ii) the catalytic decomposition of H_2O_2 under acidic pH and (iii) direct photolysis of H_2O_2 under UV light, resulting in a continuous recycling process for the Fe catalyst, as long as H_2O_2 is present in the reaction medium [1]. The photochemistry of Fe(III) is dominated by aqueous ligand-to-metal charge transfer (LMCT) processes in which the iron is reduced and $HO^\bullet$ is formed [2–5];

$$Fe^{3+} + H_2O \leftrightarrow Fe(OH)^{2+} + H^+ \quad K = 2.89 \times 10^{-3}\,M \qquad (5.1)$$

At pH = 2.8, i.e., the optimal Fenton reaction pH, the $Fe(OH)^{2+}/Fe^{3+}$ molar ratio is 1.8 [4];

$$Fe(OH)^{2+} + h\upsilon \rightarrow Fe^{2+} + HO^\bullet \text{ (possible at } 200 < \lambda < 450 \text{ nm)} \qquad (5.2)$$

$$Fe^{2+} + H_2O_2 \rightarrow Fe^{3+} + HO^\bullet + OH^- \qquad (5.3)$$

In the above given Fenton's reaction, the formation of a hydrated $Fe-H_2O_2$ complex is thermodynamically favored. Actually, the reaction consists of a ligand exchange reaction ($H_2O_2-H_2O$) in the first ligand inner-sphere of the Fe(II) cation [6]. For reasons of simplicity, the monomeric complex is shown as Fe^{2+} (aq).

$$Fe^{3+} + H_2O_2 \rightarrow Fe^{2+} + HO_2^\bullet + H^+ \quad 6.6 \times 10^{-4} \quad M^{-1}s^{-1} \quad (5.4)$$

$$Fe^{2+} + HO_2^\bullet/O_2^{-\bullet} \rightarrow Fe^{3+} + H_2O_2 \quad 1.2 \times 10^6 \quad M^{-1}s^{-1} \quad (5.5)$$

$$H_2O_2 \rightarrow 2HO^\bullet \quad (\text{at } \lambda < 300\,nm) \quad (5.6)$$

The quantum yield of Fe(III) photoreduction given in Eq. 5.2 for the $FeOH^{2+}$ complex ($\Phi Fe(OH)^{2+}$) has been reported as 0.21 under steady-state illumination using laser flash photolysis [7]. Numerous studies reported $\Phi_{\lambda,Fe(OH)}^{2+}$ in the range of 0.1–0.2 [3, 8–10]. Among them, the work by BenkelbergH and Warneck [8] reported the $\Phi_{\lambda,Fe(OH)}^{2+}$ values at every 10 nm interval over the range 280–370 nm using a monochromator. It should be noted here that secondary reactions have a noticeable effect on the yield of Fe(II) formation. In the laser photolysis experiments, the quantum yield error is minimal because the Fe(II) and Fe(III) are measured just after the laser pulse and the secondary dark reactions are not important [7].

The catalytic decomposition of H_2O_2 by Fe^{2+} and Fe^{3+} produces $HO^\bullet$ and $HO_2^\bullet$, respectively. Spectrophotometric kinetic studies have demonstrated that the reaction between Fe^{3+} and H_2O_2 primarily leads to the formation of an Fe(III)-hydroperoxy complex formulated as $Fe(III)(HO_2)^{2+}$ [4]. At very high concentrations of H_2O_2, diperoxo complexes have also been suggested [4]. The formation of these complexes is very fast and equilibria are attained within a few seconds after mixing of Fe(III) and H_2O_2 solutions. The equilibrium rate constant for the formation of $Fe(III)(HO_2)^{2+}$ is 3.1×10^{-3} at 25 °C, whereas its decomposition rate to Fe^{2+} and $HO_2^\bullet$ was determined as 2.7×10^{-3} s [11]. Considering that the overall reaction between Fe^{2+} and H_2O_2 producing $HO^\bullet$ is 63 M^{-1} s^{-1} [11], the direct Fenton reaction can be used to describe the rate of the initiation step of the catalytic H_2O_2 decomposition mechanism. The reaction has its optimum at pH 2.8–3.0, above which pH Fe^{3+} starts to precipitate; at pH values <2 or H_2O_2 concentrations less than 0.1 M, the fraction of Fe(III) in the form of Fe(III)-peroxy complexes becomes very small and the Fe^{2+} formation rate decreases. Consequently, the rate of $HO^\bullet$ formation depends of the solution pH and H_2O_2 concentration that in turn effects the steady-state Fe^{2+} concentration in the aqueous solution [4, 11].

The amount of Fe^{2+} in the reaction medium mainly depends on the actual H_2O_2 concentration; provided H_2O_2 is present, the Fe^{2+} concentration is low (only about 10% of the total Fe concentration), because all regenerated Fe^{2+} is directly consumed via H_2O_2 in the dark (thermal) Fenton process to produce new $HO^\bullet$ that initiate free radical chain reactions with the organic pollutant present in the reaction solution, eventually resulting in mineralization to the oxidation end products CO_2, H_2O and mineral acids or salts.

In industrial wastewater treatment applications it is desirable that the reaction is complete within minutes. Since the degradation rate appreciably increases with increasing light intensity, the use of high intensity UV light sources is recommended. Due to the fact that Fe iron absorbs UV, near-UV and visible light even at slightly acidic to neutral pH values (active pH range: 2–7), Photo-Fenton reactions play a key role in photochemical transformation reactions taking place in the atmosphere (water droplets) and natural waters (surface and sea water). Several studies have indicated that the $HO^\bullet$ concentration in natural waters depend on water composition (alkalinity, hardness, turbidity, etc.) as well as on the spatial and temporal variations in the solar spectrum. On the basis of several investigations, it has been suggested that the reaction between H_2O_2, a common constituent in lake and river water, and photochemically produced Fe^{2+} may be involved in the oxidation of several natural and synthetic chemicals found in freshwater. Oxidation via direct UV (near-UV) photolysis of H_2O_2 is comparatively insignificant because H_2O_2 only weakly absorbs solar irradiation [4].

Kinetic studies conducted with ferric oxalate and citrate being abundant and forming strong complexes with ferric iron, under varying pH values and the presence of oxygen (air-saturated water) and argon (de-aerated water) revealed that the $HO^\bullet$ production rate and yield is a strong function of these parameters; (i) the Fe-oxalate is the better choice in acidic pH medium and Fe-citrate system performs better at slightly acidic to neutral pH, (ii) the photoproduction rate of Fe(II) in the oxalate system is nearly the same in air- and argon-saturated solutions, (iii) the reaction with H_2O_2 is the dominant pathway for Fe^{2+} oxidation that is photochemically produced in the presence of H_2O_2 [12, 13];

$$Fe^{3+}(L-) + h\upsilon \rightarrow \ Fe^{2+} + L^\bullet \ (L = organic \ ligand) \qquad (5.7)$$

The quantum yield of this reaction is dependent on pH, irradiation wavelength and type/form of Fe(III)-hydroxo complex.

The advantages of Photo-Fenton oxidation over ozonation and heterogeneous photocatalysis have been discussed in former work and can be summarized as (i) fast homogenous reactions with no mass transfer limitations and (ii) relatively less operating costs considering electric energy requirements and chemical costs. Photo-Fenton processes are known to be suitable for the treatment of wastewaters, with high concentrations of organics, due to their high performance and competitive economy [14–18].

5.2.2 Active Oxidants Involved in the Photo-Fenton Reaction

The free radical mechanism involved in the Fenton reaction has been questioned from time to time, and alternative oxidants, whose structure was written as Fe^{4+}(aq), have been proposed to be responsible of thermal and Photo-Fenton systems based on oxidation product identification (Fig. 5.1).

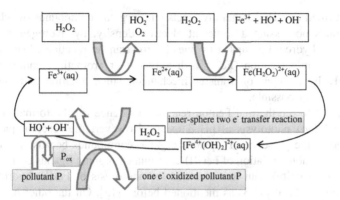

Fig. 5.1 Scheme of possible mechanisms involved in the thermal Fenton Reactions—Reprinted from Ref. [19] with kind permission of the American Chemical Society

Pignatello et al. [13] carried out competition kinetics and investigated the kinetic deuterium isotope effect (KDIE) for cyclohexane to question the existence of an additional active oxidant in the Photo-Fenton process. The slight but statistically significant difference between the Photo-Fenton and selected $HO^\bullet$-only oxidation systems (i.e., H_2O_2/UV-C) obviously speak for an additional pathway for cyclohexane loss in the Photo-Fenton reaction. The addition of the powerful $HO^\bullet$ scavenger tert-butyl alcohol to the Photo-Fenton and H_2O_2/UV-C oxidation systems at varying concentrations also supported this argument since its inhibitory effect on cyclohexane degradation was more significant for the H_2O_2/UV-C oxidation process. These obtained experimental findings are in accordance with some studies reporting KDIE values for chelated high-valent oxo-iron complexes. Besides, appreciable epoxide formation was observed during the Photo-Fenton oxidation of cyclohexene, but was not significant for dark Fenton and H_2O_2/UV processes supporting the involvement of ferryl complexes in the Photo-Fenton reaction. Alkene epoxide formation is rather uncharacteristic for $HO^\bullet$ involving oxidation reactions.

Regarding the product distributions of H_2O_2/UV and Photo-Fenton oxidation processes, compared to H_2O_2/UV, the Photo-Fenton reaction produces around 10 times more dichloroacetic acid from trichloroethane, 30 times more trichloroacetic acid from tetrachloroethene and 20 times more dichloroacetic acid from trichloroethene. This effect can be explained either by (i) reactions with iron (Fe^{2+}/Fe^{3+}) in addition to $HO^\bullet$ or (ii) by reactions with alternative oxidants (iron containing oxidants, such as ferryl), with the latter explanation being the more plausible one, since iron cannot influence $HO^\bullet$ attack unless it coordinates with $HO^\bullet$, nor can iron influence Cl atom shifts except through coordination which is rather unreasonable.

Owing to the fact that $HO^\bullet$ production might be too slow to compete with direct electron transfer between the substrate and the hydrated higher valent iron species (most likely Fe^{4+}(aq); $E_o(Fe^{3+}/Fe^{4+}) = 1.4$ eV vs NHE at pH 7; and 1.8 at pH 10.0 [4], the involvement of Fe^{4+} as a reactive intermediate of the Fenton reaction cannot be ruled out. HO$^\bullet$ generally reacts with organic compounds by addition to double bonds possessing a sufficient electron density, by hydrogen abstraction from alkyl or hydroxyl groups or by one electron transfer reactions. In contrast, the reaction of a metal cation (Fe^{4+}) with aliphatic or aromatic organics proceeds exclusively by an electron transfer mechanism since addition and abstraction reactions are not possible.

In contrast to the thermal Fenton reaction, evidence for the formation of HO$^\bullet$ during Fe^{3+} (aq) photolysis at pH 3 exists. However, the reaction pathway is suppressed in the presence of aqueous organic compounds because the quantum yields for the photooxidation of Fe(III)-coordinated organic ligands is significantly higher (around unity) than that obtained for photolysis of Fe(III) hydroxo complexes (being 0.21 at pH 3) as mentioned before [13]. On the other hand, selectivity and pulse radiolysis studies confirmed that the Photo-Fenton reactions mainly involve the intermediacy of HO$^\bullet$ rather than more selective species including ferrate [19].

5.2.3 Effect of Cl$^-$ Ions on the Photolysis of Iron Complexes

Photochemical reactions have been shown to transform Cl$^-$ into Cl$^\bullet$ that are reactive but more selective species than HO$^\bullet$ leading to the chlorination of organic compounds and resulting in the build-up of toxic and/or carcinogenic, hence undesired adsorbable organically bound halogens (RCl) collectively known as the AOX (pollution) parameter. As a consequence, the investigation of the Cl$^-$ effect on photodegradation of organic pollutants found in water and wastewater is highly warranted [7].

Fe(OH)$^{2+}$ is the dominant UV absorbing species in solution at 300–400 nm, pH 2–3 and in the absence of Cl$^-$ ions using Fe(ClO$_4$)$_3$ as the iron source in the solution [20]. On the other hand, in the presence of Cl$^-$ ions ($\sim$100 mM), the FeCl^{2+} species is the dominant UV absorbing species in solution at a pH of around 1 [7]. The reactions taking place during UV photolysis of FeCl^{2+}, and FeCl^{2+} complexes and the main reactions taking place are given below [7, 20, 21];

$$FeCl^{2+} + h\upsilon \rightarrow Fe^{2+} + Cl^\bullet \quad \Phi FeCl^{2+} = 0.5 \text{ at } 347 \text{ nm} \qquad (5.8)$$

$$FeCl^{2+} + h\upsilon \rightarrow FeCl^+ + Cl^\bullet \qquad (5.9)$$

Cl$^\bullet$ atoms instantaneously react with Cl$^-$ to give Cl$_2^{-\bullet}$;

$$Cl^\bullet + Cl^- \rightarrow Cl_2^{-\bullet} \quad 2.1 \times 10^{10} \quad M^{-1}s^{-1} \qquad (5.10)$$

The maximum absorption band of Cl$_2^{-\bullet}$ was found as 340 nm. Additional reactions that have to be considered in the presence of Cl$^-$ are given below;

$$Cl_2^{-\bullet} + H_2O_2 \rightarrow HO_2^\bullet + 2Cl^- + H^+ \quad 9 \times 10^4 \quad M^{-1}s^{-1} \qquad (5.11)$$

$$Cl^{\bullet} + H_2O_2 \rightarrow HO_2^{\bullet} + Cl^- + H^+ \quad 1 \times 10^9 \quad M^{-1}s^{-1} \tag{5.12}$$

$$Cl^{\bullet} + H_2O \rightarrow ClHO^{\bullet-} + H^+ \quad 1.3 \times 10^3 \quad M^{-1}s^{-1} \tag{5.13}$$

$$ClHO^{\bullet-} \leftrightarrow HO^{\bullet} + Cl^- \quad 6.1 \times 10^9 \, s^{-1} \tag{5.14}$$

$$Cl^{\bullet} + Fe^{2+} \rightarrow Cl^- + Fe^{3+} \quad 5.9 \times 10^9 \quad M^{-1}s^{-1} \tag{5.15}$$

$$ClHO^{\bullet-} + Fe^{2+} \rightarrow Cl^- + HO^- + Fe^{3+} \quad 1.3 \times 10^8 \quad M^{-1}s^{-1} \tag{5.16}$$

$$Cl_2^{\bullet-} + Fe^{2+} \rightarrow 2Cl^- + Fe^{3+} \quad 1.4 \times 10^7 \, M^{-1}s^{-1} \tag{5.17}$$

$$Cl_2^{\bullet-} + Cl_2^{\bullet-} \rightarrow Cl_3^- + Cl^- \quad 3.1 \times 10^9 \quad M^{-1}s^{-1} \tag{5.18}$$

The organic pollutant R reacts with $Cl_2^{-\bullet}$ to form chlorinated organics (AOX, RCl);

$$R + Cl_2^{-\bullet} \rightarrow RCl + Cl^- \quad \approx 3 \times 10^9 \quad M^{-1}s^{-1} \tag{5.19}$$

A significant decrease in the degradation rate of organic pollutants was observed in the presence of 10 mM Cl^-, but further Cl^- addition only marginally affected the degradation rate. Increased H_2O_2 concentrations (>3 mM) and the presence of light irradiation decreased the amount of AOX, RCl produced in solution as compared to dark Fenton processes. From Eqs. 5.10–5.18 it was possible to estimate $[Cl_2^{-}] \approx 10^{-7}$ M. This value is close to the concentration of $HO_2^{\bullet}$ in aqueous solution. The concentrations of $Cl^{\bullet}$ and $ClHO^{\bullet-}$ were found to be about two orders of magnitude lower than $[Cl_2^{\bullet-}]$. The $HO^{\bullet}$ are estimated to be three orders of magnitude below $[Cl_2^{\bullet-}]$ in the presence of >10 mM Cl^-.

5.2.4 Temperature Dependence of Photo-Fenton Reactions

Many parameters, such as the initial concentrations of ferrous/ferric salt and hydrogen peroxide, the molar Fe/H_2O_2 ratio, reaction pH, light intensity and temperature influence the efficiency of Photo-Fenton oxidation systems. Among these operating parameters, it has been reported that elevating the reaction temperature significantly increased the activity and kinetics of the Photo-Fenton reaction [22–26]. Using solar energy not only as the photon source, but also as the heat source appreciably enhances the treatment efficiency. For example, Sagawe et al. [25] developed an insulated solar Fenton hybrid process, which was equipped with suitable insulation and a heat exchanger, and utilized the solar heat energy for the purpose of improving the activity of the treatment system. The mechanistic interpretation for the thermal enhancement and a kinetic model that could be used to explain the temperature dependence of system activity was studied by Lee and Yoon [27] for Fenton and Photo-Fenton oxidation of p-chlorobenzoic acid.

Their study indicated that the photolysis rate of the $Fe(OH)^{2+}$ complex during Photo-Fenton oxidation initiated with ferric perchlorate was increased by about three-fold at pH 2.0, when the reaction temperature was increased from 25 to 50 °C. The enhancement of $Fe(OH)^{2+}$ photolysis could be explained by the increase of the $Fe(OH)^{2+}$ concentration and the temperature dependence of the quantum yield obtained for the photochemical reduction of Fe(III). Although elevating the temperature from 25 to 50 °C accelerated the steady-state $HO^{\bullet}$ concentrations, the enhancement was significantly higher for the dark Fenton-like oxidation system (27-fold enhancement) than in the Photo-Fenton oxidation system, resulting a three-fold enhancement. Accordingly, this study proposes the effectiveness of the high-temperature (Photo)-Fenton and Fenton-like oxidation systems in treating industrial wastewater. Furthermore, the quantitative interpretation on the temperature dependence of the Photo-Fenton oxidation system can be utilized for (photo)chemical reactor design considering heat transfer and reaction temperature. In order to fully utilize the thermal enhancement in the Photo-Fenton process the additional energy for heating up the process water is required. Hence, using solar energy as a heat source appears to be a good choice for this purpose. In that study the activation energy of $Fe(OH)^{2+}$ photo-reduction was determined as 11.4 kJ/mol (average value of 2 experiments) from the results obtained at pH 2 and 3.

5.2.5 Ferrioxalate-Photo-Fenton Systems

Photolysis of Fe(III) complexes and the subsequent Fenton reaction are defined as Photo-Fenton oxidation systems. In iron-rich environmental compartments with low DOC contents, photolysis of Fe(III)-hydroxo complexes mainly contributes to $HO^{\bullet}$ production, whereas many soil and aqueous environments being rich in iron oxides, humic and fulvic acids as well as simple caboxylic acids such as oxalate, the photolysis of Fe(III)-organo complexes may become an important source of $HO^{\bullet}$ [28].

For wastewaters that are strong UV-absorbers because of their high pollutant content and chemical oxygen demand (several thousand mg/L), the UV–visible light/ferrioxalate-Fenton process (up to $\lambda = 500$ nm) is a promising alternative to the more conventional H_2O_2/UV-C treatment process, since H_2O_2 has a relatively low extinction coefficient (18 M^{-1} cm^{-1} at 254 nm) and its UV absorption band is limited to the short-UV light region. Former studies have shown that a combination of UV light and H_2O_2 provides a potential treatment process to partially destroy/detoxify organic pollutants. The efficiency of the process is inversely proportional to the pollutant concentration and UV absorbance of the polluted water or wastewater [29]. For heavily contaminated effluents very high H_2O_2 concentrations (>1,000 mg/L) and high UV doses (strong light intensity and/or long irradiation time) are required for effective treatment. Hence, Ferrioxalate/Fenton/UV systems would offer significant advantages over H_2O_2/UV-C oxidation since ferrioxalate absorbs over a broad range of wavelengths, and thus utilizes the UV–Vis lamp output more efficiently [29]. This process is especially more efficient for waters contaminated with aromatic pollutants because these compounds form hydroxy and polyhydroxy aromatics during photocatalytic degradation with strongly UV absorbing characteristics. Hence, these oxidation intermediates compete with Fe(III) and H_2O_2 for UV and near-UV light, thus blocking the UV and near-UV light spectrum and slowing down the reaction rates appreciably. However, ferrioxalate absorbs even visible light portion up to 400–450 nm enabling the use of natural (solar) light for treatment and eliminating the competition with the polluted wastewater for UV and near-UV light [29].

Several studies have investigated the pH-dependence and the impact of different organic Fe-complex concentrations on pollutant transformation kinetics in a systematic way. $HO^{\bullet}$ probe compounds such as the banned herbicide atrazine were selected as model pollutants in these studies since atrazine is known to react conservatively except with the proposed photochemically formed $HO^{\bullet}$ [30]. It was observed that in these oxidation systems, $HO^{\bullet}$ attack was the only pathway for atrazine degradation. In the presence of oxalate, atrazine abatement rate increased at pH 3.2–4.3 and declined specially above 7.0, whereas no photocatalytic activity was observed for pH values above 7.9–8.0. In the absence of oxalate, atrazine transformation proceeded slower and only below a pH of 4.1. Results of different experimental studies indicated that pH and oxalate concentrations both control iron speciation and photochemical activity. Besides, the rate of Fenton's reaction and hence H_2O_2 and Fe(II) speciation affect the Photo-Fenton reaction. Finally it could be established that oxalate acted as a $HO^{\bullet}$ scavenger at elevated concentrations. The calculated $HO^{\bullet}$ concentration was in the range of about 10^{-12} M and hence much higher than in natural waters, e.g., in lake waters ($HO_{ss}^{\bullet} \approx 10^{-16}$ M). Considering the low Fe(III) solubility over a wide pH range, a considerable fraction of Fe(III) is in amorphous and crystalline oxide or hydroxide forms in environmental systems. Hence, it remains a pertinent question whether the Photo-Fenton reaction plays an important role in iron containing solid phases found in the environment.

As aforementioned, Fe(III) ions form strong complexes with carboxylates and polycarboxylates. These complexes are photochemically active and generate ferrous ions on irradiation according to the following reaction [28];

$$Fe(III)(RCO_2)^{2+} + h\upsilon \rightarrow Fe^{2+} + CO_2 + R^{\bullet} \qquad (5.20)$$

$R^{\bullet}$ may react with aqueous oxygen and degrade further. Since carboxylates are formed during photocatalyzed oxidation of organic pollutants, photodecarboxylation is expected to play an important role in the degradation/mineralization of organic compounds. The quantum yield of Fe(II) formation in reaction (5.20) varies with the type of carboxylate ligand (formate, maleate, oxalate, etc.) and irradiation wavelength. The ferrioxalate complex $(Fe(C_2O_4)_3)^{3-}$ is the best known and most often studied example of such type of organic complexes. Ferrioxalate is also known as an old, popular actinometer widely used to measure UV and near-UV light intensity [31]. Under acidic pH(~ 3), ferrioxalate photolysis proceeds by the following reactions;

$$Fe(C_2O_4)_3^{3-} \rightarrow Fe^{2+} + 2\ C_2O_4^{2-} + C_2O_4^{-\bullet} \qquad (5.21)$$

$$C_2O_4^{-\bullet} + Fe(C_2O_4)_3^{3-} \rightarrow Fe^{2+} + 3\ C_2O_4^{2-} + 2CO_2 \qquad (5.22)$$

$$C_2O_4^{-\bullet} \rightarrow CO_2 + CO_2^{-\bullet} \qquad (5.23)$$

$$C_2O_4^{-\bullet}/\ CO_2^{-\bullet} + \ O_2 \rightarrow 2CO_2/CO_2 + O_2^{-\bullet} \qquad (5.24)$$

The quantum yield for Fe(II) generation is 2.0, but reaction (5.24) and the reverse of reaction (5.22) reduce the actual yield. Therefore the actual measured quantum yield for Fe(II) generation from ferrioxalate is 1.0–1.2 for the wavelength range 250–450 nm and decreases with increasing wavelength. In air-saturated aqueous solutions under acidic pH (≈ 3), the oxalate radical ($C_2O_4^{-\bullet}$) reacts with molecular oxygen and generates the $HO_2^\bullet$, that disproportionates to generate H_2O_2 [28]. This H_2O_2 may react with the other in situ formed reagent Fe(II) to produce $HO^\bullet$, thus providing a continuous Fenton's reagent to the reaction medium [1].

5.2.6 Heterogeneous Fenton and Photo-Fenton Systems

Considering that Photo-Fenton is not a stand-alone treatment process and hence often designed prior to biological wastewater treatment, the pH as well as iron concentration of the Photo-Fenton-treated effluent can be a major operation problem. After homogenous Photo-Fenton treatment, the effluent pH has to be readjusted to neutral values prior to biotreatment to prevent biomass inhibition and to enhance iron removal by iron(III) hydroxide precipitation.

In the scientific literature, different alternatives methods and systems have been proposed to overcome the limitations of the Photo-Fenton treatment process; the activity and efficiency of zero-valent iron, iron oxides, and iron-loaded zeolites have been explored. Among them, alpha-Fe_2O_3, alpha-FeOOH, beta-FeOOH, gamma-FeOOH and Fe_3O_4 are iron oxide/oxihydroxide semiconductors with photocatalytic properties in the near-UV visible spectral region [32]. In aerated environments, after absorption of UV or visible light, these semiconductors enable the simultaneous reduction of adsorbed oxygen to superoxide radicals ($O_2^{-\bullet}$) and the oxidation of pollutants. These iron oxides exhibit different photocatalytic behavior and activity. Among them, gamma-FeOOH has the highest surface area (127 m^2/g) and the lowest point of zero charge (=3.9). In former studies it has been observed that ferric oxides partially dissolve to give Fe^{3+} in aqueous solution, that in turn oxidizes target pollutants in aqueous reaction solution (e.g., sulfite). Under UV or visible light, oxidation efficiencies are greatly enhanced; iron(III) oxide absorbs light of suitable energy to promote the generation of electron–hole pairs on the semiconductor surface that initiate redox reactions. Adsorbed pollutants undergo hole and free radical oxidation in the reaction medium. In addition, soluble Fe(III) complexes are also generated and undergo photoreduction depending on the wavelength of the incident light flux. It should be noted however, that photoreduction even under near-UV light plays a minor role in HO$^\bullet$ production, since the quantum yield at 360 nm is only 0.017 [32]. Several studies have indicated that the iron oxides follow the decreasing reactivity alpha-Fe$_2$O$_3$ > beta-FeOOH $\approx$ gamma-FeOOH > alpha-FeOOH. Photooxidation can be important considering that the bangaps of these iron(III) oxides range between 2.02 and 2.12 eV [33].

He et al. [34] reported the photodegradation of the azo dye Mordant Yellow in aqueous dispersions of H$_2$O$_2$/hematite (alpha-Fe$_2$O$_3$), goethite (alpha-FeOOH), and akageneite (beta-FeOOH) at neutral pH (=8) values and UV irradiation with a 100 W Hg lamp ($\geq$330 nm). The fastest degradation was obtained for the iron oxide goethite. Application of ESR spin-trapping techniques indicated the formation and intermediacy of HO$^\bullet$ in azo dye degradation. More than 75% of the dye was degraded after 400 min treatment by following first-order kinetics. The TOC of the dye solution decreased from 4.0 mg/L to 0.1 mg/L after 6 h of photocatalytic treatment. Only a minor degree of dye decomposition occurred in the absence of UV light (18% at the end of the reaction). During the reaction iron dissolution was negligible (total dissolved iron always remained below 20 μM) and hence the Fenton's reagent contribution could be neglected. The reaction mechanism was proposed as follows:

$$Fe(III)OH + H_2O_2 \rightarrow -Fe(III)O - OH + H_2O \tag{5.25}$$

$$Fe(III)O - OH + h\upsilon \rightarrow -Fe(IV) = O + HO \tag{5.26}$$

$$Fe(IV) = O + H_2O \rightarrow -FeFe(III)OH + HO \tag{5.27}$$

The reactions are initiated by a surface complex formed between H$_2$O$_2$ and the oxide surface metal centers. The surface Fe is immobilized in the crystal lattice and octahedrally coordinated by O$_2$− and OH$^-$. The excited Fe(III)O-OH coordination bond is broken to produce active Fe(IV) = O species and HO. Fe(IV) = O is unstable and immediately reacts with H$_2$O forming another HO$^\bullet$.

On the other hand, immobilization of iron on a supporting material (membranes, oxides) is a way to overcome the obstacles of the Fenton's reagent including pH dependency, pH re-adjustment requirements and solid waste (iron hydroxide) production [35, 36]. These membranes enable a reaction under neutral pH conditions. The supporting material has to be stable, inert, recyclable, a good complexing agent and transparent to UV–Vis radiation.

The efficiency of immobilizing iron ions on Nafion$^\circledR$ (reinforcing agent: Teflongrid) membranes, being a good Fe complexing agent of high stability on the degradation and mineralization of 1.4 mM 4-chlorophenol, could be demonstrated by Maletzky and Bauer [37] for an overall treatment period of 300 min at pH 7.5. The influence of membrane area and thickness as well as its reuse potential on TOC removal was also tested. Nafion$^\circledR$ membranes of different thickness (0.18 and 0.43 mm) were prepared by soaking them into a solution containing 10 mM ferrous sulfate for 24 h at room temperature and loaded with iron by employing an ion exchange method. An increase in the membrane area from 70 to 100 cm^2, and 2 × 100 cm^2 (two layer Nafion system) increased the TOC removal efficiency from 18.5 to 20.1, and 29.9%, respectively. No iron was detected in the solution bulk and the chloride concentration increased continuously over a treatment period of 180–240 min to 100% of its stoichiometry. Iron loading affected the TOC removal efficiency only after 180 min Photo-Fenton treatment, whereas membrane thickness had a significant impact on treatment results. TOC removal decreased from 50.3% (iron loading: 0.014 g iron/g membrane) to 29.9% (iron loading: 0.011 g iron/g membrane) for the same membrane area (2 × 100/200 cm^2). The Nafion reuse potential was tested as follows; after use, the membrane was rinsed and stored in distilled water at pH 6.5 for 18 h, then the membrane was used in a Photo-Fenton run, thereafter it was rinsed in distilled water at pH 1.5 for 3 h to enable reactivation. After reactivation, the membrane became colorless, and was reloaded with iron via ion exchange method. Results indicated that the membrane could be reactivated and reused efficiently. The membrane was resistant to aging, and bulk Fe concentration always remained below 2 mg/L and hence the effluent discharge limits of environmental legislations.

In another related study, the degradation of 2,4-dichlorophenol (TOC 72 mg C/L) was carried out on Fe (1.78%)—Nafion® membranes in the presence of H_2O_2 (10 mM) under 80 mW/cm^2 visible light irradiation [36] at pH values between 2.8 and 11.0. Homogeneous Photo-Fenton reactions were capable of degrading 2,4-DCP only up to pH 5.4. The Fe-Nafion membrane appeared to be effective over many cycles during the photo-catalytic degradation exhibiting a superior and stable performance without any leaching out of Fe^{3+} ions. The degradation at the surface of the Nafion-Fe membrane seemed to be controlled by mass transfer rather than by chemical reaction. The heterogeneous system used to degrade 2,4-dichlorophenol also worked efficiently for other chloro-carbons like 4-chlorophenol, 2,3-chlorophenol and 2,4,5-trichlorophenol.

5.2.7 Process Integration with Biological Treatment

Biological treatment of wastewater is often the most cost-effective alternative compared to other treatment solutions. However, since industrial effluents are typically containing difficult to treat, toxic and/or refractory organic chemicals stand-alone biotreatment systems are not effective in their removal. Hence integrated chemical, photochemical and biological treatment processes have been proposed in these cases [38]. Several studies have been developed and demonstrated the effectiveness of coupled chemical/photochemical + biological treatment methods; these include up flow anaerobic sludge blanket (anoxic) pretreatment followed by H_2O_2/UV-C or ozonation systems, or activated sludge treatment or attached growth systems (biofilters) followed by advanced oxidation processes (suspended or fixed film heterogeneous photocatalysis, Fenton's reagent, O_3 and

H_2O_2 coupled with UV, etc.). These processes have been intensively applied to different industrial effluents and pollutants at laboratory and pilot scale and promising results were obtained [26]. Among them, one of the most often investigated combinations is the sequential Photo-Fenton + activated sludge treatment processes since even visible light can promote photochemical oxidation of organics [39].

In a study conducted by Sarria et al. [40], a coil-shaped photochemical batch reactor equipped with a 400 W medium pressure Hg lamp mainly emitting at 366 nm was designed for different flow rates, organic loads, recirculation and oxidant addition rates. The photoreactor was connected to a fixed bed bioreactor colonized by sewage sludge. The probe compound to be treated was AMBI and real industrial effluent bearing AMBI (5-amino-6-methyl-2-benzimidazolone), a commercial azo dye intermediate. The bioreactor was regularly fed with nutrient and buffer solutions. The technical feasibility of the integrated photochemical + biological treatment system was tested in a parabolic collector equipped with three modules (collector, photoreactor and total reactor including pipelines and connections) and irradiated with solar light. In the studied pH range (2–4), Fe(III) was present in at least four different Fe(III) hydroxocomplexes. It was observed that the Fe(III)-AMBI complex formed was thermally stable. After 300 min treatment 90% AMBI and 30% DOC were removed. AMBI abatement started immediately whereas DOC removal was delayed to the formation of oxidized intermediates that usually have a lower reactivity toward $HO^•$ than the original compounds. Moreover, the photodegradation of AMBI was significantly inhibited in the presence of isopropanol, a well-known free radical scavenger. Besides, oxygen played an important role in the Fe-assisted photocatalytic reactions since oxygen reacts with organic radicals to give products containing hydroxyl and carboxyl functional groups that form photoactive Fe(III) complexes. Oxygen also participates in the generation of H_2O_2 and the mechanism involved is reduction of molecular oxygen to $O_2^{•-}$ radicals and its conjugate acid, the $HO_2^•$, leading to their disproportionation to H_2O_2 and O_2. According to the Zahn-Wellens test employed to evaluate ultimate biodegradability of the aqueous pollutant, the residual DOC contains relatively biodegradable ingredients after total degradation of AMBI via Photo-Fenton reaction. In order to optimize the integrated treatment system, biodegradation was coupled with the photochemical treatment system that was operated under varying conditions (0–600 min). In order to obtain a compromise between the highest overall treatment efficiency of the coupled system and the shortest photochemical pretreatment, 300 min photochemical treatment was applied since biotreatment was negatively affected when increasing the photochemical treatment time beyond 300 min. On the other hand, shorter photochemical treatment times were not enough to eliminate the originally refractory compound AMBI. Real industrial wastewater containing AMBI as the principal component was also subjected to integrated photochemical + biological treatment. Considering the environmental limitations for iron in an effluent treatment that goes on to be discharged into natural waters (2 mg/L) or to a biological wastewater treatment plant (i.e., municipal sewage treatment works 20 mg/L), the concentration range of

Fe(III) to be used was in the range of 2.0–16.8 mg/L. When the concentration of Fe(III) was increased, the individual and overall performances of the coupled system was almost ten times higher. The addition of H_2O_2 increased the mineralization rate from 36 to 65% DOC removal after 5 min photodegradation, however for both cases the same overall removal efficiency of around 90% (65 + 27% versus 36 + 55% DOC removal) was achieved in the coupled system. In other words, a higher biological efficiency was observed in the $Fe/O_2/UV$ light system than in the Photo-Fenton oxidation process. In order to assess the practical application potential of the solar Photo-Fenton reaction as a pretreatment step prior to biodegradation, the treatment cost of this technology using a full-scale integrated system of 500 m^2 was calculated by making several technical considerations. The total cost estimation was done by using an interest rate of 13% over a 15-year period. The photochemical pretreatment cost was estimated to be 22 USD per m^3 of AMBI-containing wastewater (organic carbon: 4,000 mg/L COD). After pretreatment, it is possible to send the effluent toward an aerobic biological, relatively low-cost treatment unit for full oxidation.

5.3 Oxidation of Alkyl Phenols and Bisphenol A by Photo-Fenton Process

5.3.1 Alkyl Phenols

Alkyl phenol ethoxylates (APEOs) are reported to comprise approximately 6% of the total surfactant production in the world [41, 42] and are used as cleaning products ($AP_{10-12}EOs$), detergents, emulsifiers ($AP_{n>20}EOs$), wetting agents ($AP_{8-9}EOs$) in the household, dispersing agents in industry, herbicides and pesticides ($AP_{n>15}EOs$) in agriculture [42–44]. The annual worldwide APEO production is around 500,000 tons [45]. Of the various APEOs, nonylphenol ethoxylates (NPEOs) represent approximately 80% of the total worldwide production [45, 46] whereas octylphenol ethoxylates (OPEOs) represent most of the remaining 20% [45, 46]. APEOs enter the environment primarily via landfill leachates, industrial effluents and domestic and/or industrial wastewater treatment plant effluents (dissolved in sludge), but also by direct discharge [45, 47–49, 50]. The composition of the mixture can differ considerably among the various effluents, depending on the source, type and degree of treatment [48]. The concentrations of alcyl phenols (APs) found in river and surface waters range from non-detectable to 158 µg/L [45]. Textile industry effluents, which are the major source of APEOs, contain very high levels of NP and NPEOs (2.68–180 µg/L) and ultimate wastewater treatment plant discharge contains NP in the range of <0.02 to 62.1 µg/L [51, 52].

There are conflicting reports in the scientific literature on the biodegradability of APEOs. APEOs are not readily biodegradable using standard tests, but are inherently biodegradable [53, 54]. It has been demonstrated that after a period of acclimation, APEOs, in sewage sludge or the natural environment, can undergo biodegradation to shorter-chain APEOs involving stepwise loss of ethoxy groups to lower APEOs congeners [55] and ultimately to metabolites namely alkyl phenols (APs). The microbial breakdown of NPEOs and OPEOs gives nonylphenol (NP) and octylphenol (OP), respectively, which are more hydrophobic, toxic, persistent and/or estrogenic than the parent compounds [45, 56]. NP and OP are not readily biodegradable and their ultimate elimination takes months or even longer in surface waters, soils and sediments, where they tend to be immobilised [57]. Conventional treatment methods, such us the coagulation–flocculation followed by aerobic and/or anaerobic biological processes, are not efficient enough for complete degradation of these persistent compounds. The employment of more efficient processes for removal and/or improvement of the biodegradability of such recalcitrant and disrupting species has become more urgent and necessary.

Numerous studies have demonstrated the ability of APEOs and their breakdown products, APs, to disrupt the normal function of the endocrine system of mammalians, fishes, and amphibians [51, 58–60] causing effects on uterus-, testes-, prostate-weight or other sex organ weights, effects on sperm development, vaginal opening, imposex, effects on thyroid hormone levels or synthesis and neuroendocrine pituitary effects [61]. The available acute and chronic toxicity data of NP to aquatic organisms indicates NP is highly toxic to fish, aquatic invertebrates and aquatic plants. The 28-day no observed effect concentration (NOEC) of 4-NP (4-nonylphenol) for fish ranges from 0.05 to 0.07 mg/L and the 28-day lowest observed effect concentration (LOEC) ranges from 0.12 to 0.19 mg/L [53, 54, 62]. For 4-OP (4-tert-octylphenol) chronic LOEC and NOEC were found as 0.0061 mg/L and 0.01 mg/L, respectively from the 60 day post-hatch early life stage toxicity study with rainbow trout (*Oncorhynchus mykiss*) [63]. For NPEOs, toxicity to aquatic organisms tends to decrease with increasing degree of ethoxylation. For example, acute toxicity to killifish was 1.4, 3.0, 5.4, 12.0 and 110.0 mg/L for NP, NP1EO (i.e., NPE with one ethoxylate group), NP6.4EO (i.e., NPE mixture with an average of 6.4 ethoxylate groups), NP9EO and NP16.6EO, respectively [53]. In vitro studies showed that NP and OP could activate the estrogen receptor with a potency of $\approx 7,000$ and $\approx 1,500$ times lower than that of 17β-estradiol, respectively [45, 64]. Although these substances are less estrogenic than 17β-estradiol, their concentrations in surface waters can be orders of magnitude higher than those of natural and synthetic hormones [65]. Furthermore, in many surface waters, APEOs and/or APs are present in combination with other endocrine disrupting compounds (EDCs), so that overall toxicological effects are difficult to anticipate [66]. Because of their high aquatic toxicity, several countries have had a voluntary agreement with industry not to use NP, NPEOs or OP in domestic detergents; however APEOs are still being used in several industrial applications where they cannot be replaced yet by other alternative chemicals due to several technical as well as economic reasons. OP is included in the OSPAR 1998 List of Candidate Substances (cf. List 6 in Annex 3 of the OSPAR Strategy with regard to Hazardous Substances) while NP was already identified for priority action in 1998, when the Commission adopted the OSPAR Strategy with regard to Hazardous Substances (cf. Annex 2 of this strategy). NPEOs used in cleaning agents were recommended to be phased out in 1995 for domestic use and in 2000 for industrial use under the OSPAR Convention and it is also listed as a substance for priority action on its control under the Helsinki Convention on the Protection of the Marine Environment of the Baltic Sea Area (2000).

Because of the toxic and endocrine disrupting effect of APEOs and APs, a need for enhanced technologies that can render complete or considerable destruction of these compounds in the environment has become evident. As aforementioned, AOPs are innovative technologies for water and wastewater treatment that allow total or partial elimination of persistent, toxic and/or endocrine disrupting compounds. Among them, H_2O_2/UV-C, Photo-Fenton, and TiO_2-mediated heterogeneous photocatalysis are the most investigated ones [59, 67–69]. In this section, brief information concerning laboratory studies on H_2O_2/UV-C, Photo-Fenton, Fenton oxidation and direct UV-C photolysis, of APEOs and APs, is presented to gain up-to-date information on aspects related to the influence of operating variables such as oxidant dose, UV intensity, pH, temperature, presence of free radical scavengers, etc., degree of degradation, reaction kinetics, identity and characteristics of oxidation by-products, and impact on toxicity and biodegradability, where this information is available.

There are limited source of scientific literature on direct UV photolysis of APEOs and APs. Neamtu and Frimmel [51] explored the decomposition of NP under a solar simulator equipped with a Xe lamp in the presence and absence of H_2O_2, qualitatively establishing the effects of pH, dissolved organic matter (DOM), presence of Fe(III), and H_2O_2 concentration on the reaction rate. The results indicated that the oxidation rate increases in the presence of H_2O_2, Fe(III) and DOM with DOC concentrations not higher than 3 mg/L. They found that the pseudo-first-order rate constant for NP disappearance increased with pH elevation(s) as a consequence of the larger photoreactivity of the deprotonated molecule; but it decreased with elevation(s) of initial NP concentration. Under specific conditions, phenol and 1,4-dihydroxybenzone were recognizable reaction intermediates. Addition of bicarbonate and nitrate ions retarded and enhanced the rate of degradation upon excess consumption and production of $HO^\bullet$ (by NO_3^- photolysis), respectively. Importantly, the yeast estrogen screen (YES) test showed a decrease in estrogenic activity of NP after 8 h irradiation in the presence of H_2O_2. Another study focusing on the photolysis of OP was investigated using a solar simulator in the absence/presence of dissolved natural organic matter (DNOM), HCO_3^-, NO_3^- and Fe(III) ions [48]. The effects of different operating parameters such as initial pH, initial concentration of substrate, temperature and the presence of H_2O_2 on photodegradation of OP in aqueous solution have been assessed. The present study has shown that the degradation of OP in aqueous solution by photosensitized oxidation using UV-solar simulating light is slow. The degradation rate constant for the direct photolysis of OP in DNOM-free water has been calculated to be $1.70 \times 10^{-2} h^{-1}$. Based upon the YES test results, the estrogenic activity of OP aqueous solution after 8 h of irradiation remained approximately constant for experiments conducted using the UV-solar simulator in the presence of carbonate ions and it practically disappeared when in the initial solution 50 mM H_2O_2 were added. In conclusion, the results confirm again that OP is a persistent intermediate in OPEOs degradation. The pH, initial H_2O_2 concentration, effect of DNOM and common water constituents, influenced the photodegradation of OP. The results indicated that the oxidation rate increases in the presence of H_2O_2 and NO_3^-. Phenol, 1,4-dihydroxybenzene and 1,4-benzoquinone were found as the dominant intermediate products of photodegradation of OP. Chen et al. [66] investigated the performance of a H_2O_2/UV process for the removal of a variety of phenolic and steroidal estrogens including NP. The efficacy of the process was assessed by monitoring estrogenic activity using an in vitro yeast screen assay and an in vivo fish assay. Kinetics of estrogenic activity removal in both assays followed pseudo-first-order rate law regardless of the water matrix, but the rate was faster in deionized water than in river water. The presence of bulk organics decreased the oxidation rates of the target contaminants, suggesting that these compounds acted as free radical scavengers. Another study focusing on the effects of frequently used textile preparation chemicals and common ions on the H_2O_2/UV-C treatment of a ten-fold ethoxylated nonylphenol (NPEO-10) was conducted by Arslan-Alaton et al. [70]. Within the scope of this experimental work, the effect of organic additives (two common phosphonic acid-based sequestering agents namely diethylene triamine penta-methylene phosphonic acid and 1-hydroxye-thylidene-1,1-diphosphonic acid), chloride, soda ash carbonate and the binary impact of the chloride-carbonate scavengers on the H_2O_2/UV-C treatment of NPEO-10 under acidic and alkaline pH conditions was examined. Photochemical treatment efficiencies and NPEO-10 degradation rates were evaluated in terms of NPEO-10 (the target pollutant), COD and TOC abatements as well as in H_2O_2 consumption rates and pH evolutions. Arslan-Alaton et al. [70] found that NPEO-10 (210 mg/L), could be reduced less than 10% accompanied with no changes in the COD and TOC parameters by direct UV-C photolysis in the absence of H_2O_2. The authors concluded that NPEO oxidation primarily involved a reaction mechanism with $HO^\bullet$. The experimental findings have demonstrated that photochemical treatment of NPEO-10 with the H_2O_2/UV-C process (experimental conditions: 210 mg/L NPEO corresponding to 450 mg/L COD; 30 mM H_2O_2 at an initial pH of 10.5) occurred very fast and resulted in the complete degradation of NPEO-10 to its mineralization end products. NPEO-10, COD and TOC abatements followed first-order kinetics and rate constants were calculated as 0.2211, 0.0255, and 0.0142 min^{-1}, respectively. Among the studied textile preparation chemicals and hydroxyl radical scavengers, the decreasing order of hydroxyl radical scavenging capacity was established as diethylene triamine penta-methylene phosphonic acid > 1-hydroxy ethylidene-1,1-diphosphonic acid > soda ash carbonate at pH 10.5 > chloride at pH 3.5 > chloride at pH 10.5.

The study by Rojas et al. [65] is about to characterize the mechanism and kinetics of H_2O_2/UV oxidation and Fenton's reaction to decompose aqueous-phase NP. The authors examined the dependence of transformation kinetics on wavelength in the range 230 nm $\leq \lambda \leq$ 270 nm and the concentrations of well-known $HO^\bullet$ scavengers (isopropanol or ethanol) by using monochromatic light. The rate constant for reaction between NP and $HO^\bullet$ was estimated as $1.33 \pm 0.14 \times 10^{10}$ M^{-1} s^{-1} from H_2O_2/UV photolysis experiments, a value close to those of structurally similar alkylphenols. Excess H_2O_2 (>25 mM) decreased the rate of NP decomposition as a result of radical scavenging, as predicted. Recently, a study devoted to investigate the applicability of different photochemical AOPs, namely, direct UV-C photolysis, H_2O_2/UV-C and Photo-Fenton processes (UV-A/H_2O_2/ Fe^{2+}), for the degradation of nonylphenol ethoxylate-9 (NPEO-9) in water was conducted by de la Fuente et al. [59]. In this experimental study the initial NPEO-9 concentration and pH was 300 mg/L (0.48 mM) and 6.0, respectively. In H_2O_2/ UV-C experiments, two initial NPEO-9/H_2O_2 molar ratios were tested, 1:1.0 and 1:0.5. In Photo-Fenton experiments, three initial NPEO-9/H_2O_2/Fe^{2+} molar ratios were chosen, namely, 1.0:1.0:0.1, 1.0:1.0:0.5, and 1.0:2.0:0.5; in these cases, the reaction pH was adjusted to 2.8. For dark Fenton oxidation experiments 1.0:1.0:0.5 NPEO-9/H_2O_2/Fe^{2+} molar ratio was used. The study showed that under the experimental conditions employed, NPEO-9 could be very well degraded by UV-C light alone (reaching 75% depletion in 180 min), H_2O_2 addition increased the rate, but no significant differences were found between the two studied H_2O_2 concentrations. By Photo-Fenton oxidation almost total NPEO-9 degradation was achieved in 180 min, showing a better efficiency than reactions under UV-C light. The dark Fenton process was also effective, but resulted in lower NPEO-9 removal efficiency (around 60%). The highest total TOC removal was achieved by the Photo-Fenton process and an increase in Fe^{2+} concentration increased final TOC reduction, but in contrast, a higher H_2O_2 amount did not lead to a further improvement. The initial rate constants (k_{in}, calculated by assuming pseudo-first-order kinetics up to 5 min of reaction) and the initial photonic efficiencies (ξ_{in}, calculated as the ratio of the initial rate of NPEO-9 decay to the rate of incident photons of monochromatic light on the reactor) were also calculated. The highest k_{in} and ξ_{in} values were found for Photo-Fenton oxidation at the conditions of 1:2:0.5 as 0.067 min^{-1} and 7.78%, respectively. They concluded that the Photo-Fenton processes was the optimal processes because of high NPEO-9 and TOC removal, together with a reasonably high photonic efficiency, the poor formation of toxic aldehydes and the lack of 4-NP formation.

Summary of treatment performances, reaction conditions and identified oxidation intermediates for APEOs and APs removal by AOPs reviewed in this section is given in Table 5.1. It should be also mentioned here that identified products of NPEOs or NP degradation formed in AOP treatment include NPEO with a smaller number of EO groups, carboxylic acids predominated by acetate, oxalate and formate and aldehydes predominated by formaldehyde, acetaldehyde, glyoxal and methylglyoxal [71–73].

Table 5.1 Summary of reaction conditions and identified oxidation intermediates for APEOs and APs removal by AOPs reviewed

Reference	Reaction conditions	Identified oxidation intermediates
Neamtu and Frimmel [51]	NP_o = 25 μM; 1000 W Xe short-arc lamp; H_2O_2 = 10, 20, 50 mM; pH = 5.4, 8.5; HCO_3^- = 725 mg/L; NO_3^- = 61 mg/L	Phenol; 1,4-dihydroxybenzone; 1,4-benzoquinone
Chen et al. [66]	NP_o = 40 μg/L; 15 W low pressure Hg UV lamps (253.7 nm); fluence = 0–2000 mJ/cm^{-2}; H_2O_2 = 10 mg/L	
Neamtu et al. [48]	OP_o = 15–25 μM; 1000 W Xe short-arc lamp; the photon flow in the UV range (290≤ λ ≤400 nm) = 5.68 × 10^{-7} einstein s^{-1}; H_2O_2 = 10, 20, 50 mM; pH = 6.58, 8.0; T = 15–25 °C; HCO_3^- = 725 mg/L; NO_3^- = 61 mg/L; Fe^{3+} = 100 μg/L	Phenol; 1,4-dihydroxylbenzene; 1,4-benzoquinone
Rojas et al. [65]	NP_o = 10–25 μM; 1000 W Xe arc lamp; H_2O_2 = 25–250 mM; pH = 1.8–2.7; Fe^{3+}_o = 0.135–0.300 μM	
de la Fuente et al. [59]	NPEO-9 = 0.48 mM; pH_o = 2.8 and 6.0; NPEO-9/H_2O_2 = 1:1 and 1.0:0.5 M/M; NPEO-9/H_2O_2/Fe^{2+} = 1.0:1.0:0.1, 1.0:1.0:0.5, and 1.0:2.0:0.5 (molar concentrations)	
Arslan-Alaton et al. [70]	NP_o = 210 mg/L; H_2O_2 = 30 mM; pH = 3.5, 10.5; Cl^- = 3 g/L; CO_3^{2-} = 1–5 g/L; DTPMP = 0.5–2.5 g/L; HEDP = 0.5–1.5 g/L	

5.3.2 Bisphenol A

Bisphenol A (BPA) (2,2-bis(4-hydroxyphenyl)propane, BPA), a synthetic estrogen used to harden polycarbonate plastics and epoxy resin, is the focus of a growing number of research studies and legislative actions. An estimated 6 billion pounds of BPA are produced globally annually, generating about $6 billion in sales [74]. It is fabricated into thousands of products made of hard, clear polycarbonate plastics and tough epoxy resins, including safety equipment, eyeglasses, computer and cell phone casings, water and beverage bottles. In addition, BPA is consumed as a resin in dental fillings, as coatings on cans, as powder paints and as additives in thermal paper [75–77].

Although the vast majority of BPA is converted into products, the primary sources of environmental release of BPA are expected to be effluents and emissions from its manufacturing facilities and facilities which manufacture epoxy, polycarbonate, and polysulfone resins. Any residual, unreacted BPA remaining in polycarbonate products and epoxy resins (incomplete polymerization of BPA) can leach out into food or the environment. Polycarbonate is generally stable, but some

BPA can be released from polycarbonate when it is exposed to strongly basic conditions, UV light or high heat. Epoxy resins made with BPA are stable; only residual BPA is expected to be released from epoxy resins [42, 78, 79]. Numerous publications have reported measured concentrations of BPA in streams and rivers in Japan, Europe and the United States. The reported BPA concentrations in surface waters are ranged between 0.016 and 0.5 µg/L [80]. A Japanese study reported detectable BPA in 67 of 124 water samples selected from "Water Quality Monitoring" sites for downstream rivers. The median concentration of BPA was 0.01 µg/L and 95% of the samples contained less than 0.24 µg/L BPA [81]. The concentration of BPA in the natural aquatic environment is in µg/L range; however, concentrations >10 mg/L BPA were measured in waste landfill leachates [42, 82].

Biodegradation plays a major role in the removal of BPA from the environment. Rapid and extensive breakdown of BPA has been demonstrated in a variety of laboratory biodegradation tests [83]. Based on the results of standard laboratory biodegradation tests recommended by the Organization for Economic Cooperation and Development (OECD) and from the biodegradation studies reported BPA is classified as readily biodegradable [84]. For instance, 92–98% removal was reported in the most common type of sewage treatment system, an activated sludge plant [77]. The trace amounts of BPA remaining in treated wastewater will continue to biodegrade in receiving waters and downstream of treatment plants [85, 86].

The most probable routes of human exposure to BPA are dietary intake (e.g., migration from food packaging and from repeat-use polycarbonate containers, such as baby bottles), environmental media (ambient air, indoor air, drinking water, soil and dust), use of consumer products, transport or packaging of this compound or use of epoxy powder paints [79]. Trace amounts of BPA can also enter the food chain through resin coatings used in food packaging, such as food and drink cans coated with epoxy resin lacquers. Given the tendency of toddlers to put inappropriate objects into their mouths, there is some minor potential for children to be exposed to BPA through their mouthing or accidental ingestion of consumer products. BPA was first recognized as a potential estrogen-mimicking substance more than 40 years ago. Estrogen is an important hormone in humans and other animals, controlling fertility and sexual development. Estrogenic chemicals have been implicated in incidences of abnormal sexual development in fish and other aquatic animals and possibly decreases in male fertility in humans. Although its estrogenic potency is much lower than 17β-estradiol, BPA is classed as a weak estrogen by in vitro testing. The estrogenic potency in vitro is 10^{-3} to 10^{-4} relative to that of estradiol, and 4×10^{-4} to 10^{-5} in vivo relative to that of diethylstilbestrol [87, 88]. Even at low-dose exposure, BPA induces feminization during the gonadal ontogeny of fishes, reptiles and birds. Adult exposure to environmental concentrations of BPA is detrimental to spermatogenetic endpoints and stimulates vitellogenin synthesis in model species of fish [89]. In general, studies have shown that BPA can affect growth, reproduction and development in aquatic organisms. Among freshwater organisms, fish appear to be the most sensitive species. Evidence of endocrine-related effects in fish, aquatic invertebrates, amphibians and reptiles has been reported at environmentally relevant exposure levels lower than those required for acute toxicity. There is a widespread variation in reported values for endocrine-related effects, but many fall in the range of 1 $\mu g/L^{-1}$ mg/L [90].

BPA has been evaluated as a chemical of potential concern by some U.S. agencies and other countries since the early 1980s. In 2009, the EPA Office of Water considered BPA during its development of the third Candidate Contaminant List of substances that might be appropriate candidates for future regulation. Although BPA appeared on the potential candidate contaminant list used during the screening process, BPA did not meet the combined screening criteria of potential to occur in public water systems and potential for public health concern because its measured presence in water was third Candidate Contaminant List [91]. Numerous other governmental bodies and review panels have also conducted human health risk assessments for BPA in the recent past. Japan [92], the European Union [93], and the European Food Safety Administration [94] all concluded within the past three years that the novel studies indicating low-dose, endocrine-related effects were insufficient for the purposes of hazard evaluation/risk assessment [79].

According to EPA, "BPA is an exogenous agent that interfaces with synthesis, secretion, transport, binding, action or elimination of natural hormones in the body that are responsible for the maintenance of homeostasis, reproduction or behavior". Such concerns have heightened the need for novel and advanced remediation techniques to effectively remove BPA from a variety of contaminated environmental media including water, wastewater, wastewater sludge, sediments and soils [95]. Abiotic degradation of BPA under laboratory-scale conditions has been carried out by numerous methods, including solar irradiation [96, 97], Fenton [98] and Photo-Fenton processes [99] and $H_2O_2/UV-C$ treatment [51, 96, 100].

In the following section the results of published studies concerning the degradation of BPA by direct UV photolysis, $H_2O_2/UV-C$, Photo-Fenton and Fenton processes were summarized and information on the extent of degradation, kinetic rates constants, by-product formation and impact on estrogenicity activity were discussed.

UV photolysis has been one of the most widely investigated treatment process for BPA destruction based on the facts that UV irradiation is a common practice in drinking water treatment, and photolysis is the main abiotic degradation pathway of organic matter in natural waters [87]. Rosenfeldt and Linden [100] studied the degradability of BPA by low and medium pressure mercury UV lamps and reported 5 and 10–25% reduction by direct photolysis, respectively. However, exposure of the same samples to H_2O_2/UV at a fluence rate of 1,000 mJ/cm^2 resulted in more than 90% BPA destruction regardless of the UV source. They also recorded nearly 100% reduction in estrogenic activity by H_2O_2/UV treatment. A similar study reported by Chen et al. [96] indicated that a low pressure UV lamp even when operated at 5,000 mJ/cm^2 is ineffective alone, while the efficacy is improved to 80 and 78% for BPA and estrogenic activity removal, respectively, in the presence of 10 mg/L H_2O_2. In the same study when the H_2O_2 concentration was increased to 25 mg/L no further enhancement was obtained (97% BPA and estrogenic activity removal).

Neamtu and Frimmel [101] investigated the degradation of BPA under 254 nm irradiation in different water matrices and its effect on yeast cells. They reported that the degradation and reduction in estrogenicity activity of BPA by 254 nm UV irradiation was more efficient in purified or surface water than in sewage water due to the presence of competing substances in the latter. They also observed a decline of EST in both matrices with increased irradiation time. The effect of H_2O_2 concentrations on the photooxidation of 520 μM BPA was examined in pure water, surface water and wastewater. The calculated first-order decay constant in pure water was determined as 5.0×10^{-4} s^{-1} in the presence of 500 μmol/L H_2O_2, which is about seven times higher than that without H_2O_2 addition. The HO$^{•}$, produced from UV photolysis of H_2O_2, were suggested to enhance the oxidation rate of BPA. However, at a ratio of 1.5 $[H_2O_2]/[BPA]$, the oxidation rate decreased; due to the formation of less reactive HO$_2$$^{•}$ by consuming highly reactive HO$^{•}$ at such high H_2O_2 concentration. Intermediate products of the oxidation of BPA were found to be phenol, 1,4-dihydroxylbenzene, and 1,4-benzoquinone. Moreover, estrogen activity of BPA decreased upon UV irradiation. In another study related to the impact of water composition, Zhan et al. [102] reported that the degradation of BPA in the presence of natural humic substances by solar radiation was faster than in pure water. Consequently, they suggested a mechanism to the decay process that involved the (i) production of excited BPA molecules leading to direct photolysis; and (ii) production of HO$^{•}$ from photoreactive components of humic substances leading to indirect photolysis. The impact of water matrix was further investigated by Zhou et al. [103] in synthetic samples containing BPA and ferric oxalate (Fe–Ox) complexes exposed to a high pressure Hg lamp. Based on the phenomenon that Fe-Ox complexes in atmospheric waters produce H_2O_2, which react with Fe(II) to generate HO$^{•}$ according to Fenton reaction, they found that maximum efficiency in their system was obtained at pH = 3.5 in a Fe/Ox molar ratio of 10:120. They also reported that total mineralization (160 min exposure) in these conditions was 24% as opposed to 7.3% mineralization in the absence of oxalate. The potential of Fenton reaction in UV photolysis of BPA was also examined by Katsumata et al. [99] using a Xe lamp emitting UV light at <300 nm. The study showed that maximum destruction could be accomplished at pH = 3.5–4.0 with a H_2O_2–Fe^{2+} ratio of 10 (by M), at which 50% mineralization was possible in 24 h. The optimal molar ratio of H_2O_2/Fe^{2+}/BPA for complete degradation of BPA was reported as 9.0:0.9:1.0.

The photodegradation of BPA was conducted in simulated lake water containing Fe^{3+}, algae (*Chlorella vulgaris*) and humic acid by Peng et al. [97]. Photodegradation efficiency of BPA was 36% after 4 h irradiation in the presence of 6.5×10^9 cells/L raw *Chlorella vulgaris*, 4 mg/L humic acid and 20 μM Fe^{3+}. The authors concluded that algae may produce some secretions after heavy irradiation and these secretions can produce HO$^{•}$, which can enhance the degradation of pollution. The photodegradation of BPA in aqueous solution containing raw algae, humic acid and/or Fe^{3+} was obviously greater than that in aqueous solution only containing raw algae. The enhancement rates suggest the facilitation of the formation of reactive HO$^{•}$ species through the formation of Fe^{3+} complexes with humic acid and algae. The photodegradation efficiency of BPA increased slowly with increase in concentration of Fe^{3+}. The author stated that this acceleration may be related to the formation of complexes of Fe^{3+} with humic acid and algae, which are not so preferable compared to inclination of Fe^{3+} toward a colloidal state. The photodegradation efficiency of BPA was at a maximum at 6.5×10^9 cell/L when varying the concentration of *Chlorella vulgaris* in the range of 6.5×10^9 to 17.0×10^9 cell/L in aqueous solution. The presence of humic acid also increased the photodegradation rate of BPA, but high concentrations (6 mg/L) decreased the efficiency. This is due to the reaction of $HO^\bullet$ with humic acid, which will compete with the oxidation reaction of BPA with $HO^\bullet$ at a concentration of 6 mg/L.

The effect of properties of varying iron oxides (Magh-300, Hem-420, and Hem-550) and oxalate on the degradation of BPA under UV illumination ($\lambda = 365$ nm) was examined by Li et al. [104]. The experimental results confirmed that the existence of oxalate can greatly enhance BPA degradation reaction on the iron oxides in aqueous solution compared to iron oxides alone. The properties of iron oxides strongly influenced the dependence of the BPA degradation on the oxalate concentration. The optimal initial concentrations of oxalate were determined to be 2.4, 2.4 and 0.6 mM for Magh-300, Hem-420 and Hem-550, respectively. The optimal pH value was found to be 3.93, 3.64 and 3.61, for Magh-300, Hem-420 and Hem-550, respectively. The rate of BPA degradation on different iron oxides with oxalate under UV illumination could be ranked as Magh-300 > Hem-420 > Hem-550. Furthermore, it was found that the dependence of BPA degradation is also attributable to the interaction between iron oxide and oxalate, and the formation of dissolved Fe in the solution or adsorbed Fe-oxalate species on the surface of iron oxides. Liu et al. [105] studied the photodegradation of BPA in montmorillonite KSF suspended solutions using UV–Vis irradiation (metal halide lamp $\lambda \geq 365$ nm). The effects of dissolved oxygen and carboxylates (oxalate, citrate, and pyruvate) on the photodegradation of BPA were examined. It was found that BPA could be effectively photodegraded in suspension. The BPA photodegradation was dependent on the pH of the solution and dosage of clay minerals, and it was more effective for BPA to be degraded at pH 4; the degradation rate of BPA increased with the concentration of clays in the range of 0.5–5 g/L. Results of this study suggested that both dissolved oxygen and carboxylates enhanced the oxidation of BPA. Furthermore, the acceleration in rates in case of carboxylates followed the order: pyruvate > oxalate > citrate.

In the study of Rodríguez et al. [106] the efficiencies of different solar oxidation processes on the degradation of BPA in water were evaluated. The processes tested were Fe(III) photolysis, Fe(III)-carboxylate photolysis, TiO_2-mediated heterogeneous photocatalysis, H_2O_2 photolysis and combinations thereof. The influence of pH and the presence of hematite (α-Fe_2O_3) on their efficiency, as well as the nature of intermediates formed and development of toxicity were assesed. Under the experimental conditions of this study, at pH 3 Fenton and Photo-Fenton systems were the most effective in BPA and phenolic intermediates degradation. At this pH, the highest mineralization efficiencywas achieved by Photo-Fenton and TiO_2/Fe(III) systems. At pH 6.5, the highest BPA degradation rate was found for the Fe(III)/oxalic acid/UV, and the TiO_2/UV system (either in absence or presence of

α-Fe_2O_3) was the most effective in phenolic intermediates oxidation, and α-Fe_2O_3 exerted a positive effect mainly on mineralization.

A comparative study of ultrasonic cavitation and Fenton's reagent for BPA degradation in deionised and natural waters was conducted by Torres et al. [107]. In the two cases, BPA concentration (118 µM) was under the detection limit (0.002 µM) after 90 min treatment. COD and TOC evolutions recealed that both treatment processes resulted in the formation of more oxidised intermediates with a significant decrease of COD) which are hardly mineralized as observed by a low decrease of TOC. The Fenton process exhibited better performance in the mineralization of BPA. After 180 min, 75% of COD and 20% of TOC were removed with the Fenton process. In another study conducted by Ioan et al. [98] the degradation rate of BPA in aqueous solution was investigated by the Fenton and sonolyticFenton treatment. The authors concluded that the degradation rate was strongly affected by the pH value and the initial concentrations of Fe^{2+}. Complete degradation of BPA was achieved after 60 min under both sono-Fenton and Fenton conditions; however, the sono-Fenton enhanced the degradation rate as compared to Fenton only. In the experimental study, two molar ratios of $H_2O_2/Fe^{2+}/BPA$ were used, namely 2.8:1.0:10.0 and 5:1:18, the former ratio being better among the two ratios.

A more recent contribution of Poerschmann et al. [108] focused on examining whether and to what extent aromatic intermediates with sizes larger than that of BPA, i.e., with two aromatic rings linked together by an alkyl moiety, were formed in sub-stoichiometric Fenton reactions. In this framework, distinct sub-stoichiometric $[H_2O_2]_0$:$[BPA]_0$ molar ratios of 2.3 and 4.6 ($[BPA]_0 = 44$ µM) were used. The identification of oxidation intermediates was carried out by GC/MS analysis of trimethyl-silyl ethers. The occurrence of aromatic intermediates larger than BPA, which typically share either a biphenyl- or a diphenylether structure, were explained by oxidative coupling reactions of stabilized free radicals or by the addition of organoradicals (organocations) onto BPA molecules or benzenediols. The hydroxycyclohexadienyl radical of BPA was recognized to play a central role in the degradation pathways. Ring opening products, including lactic, acetic and dicarboxylic acids, were also detected in addition to aromatic intermediates. The author concluded that oxidation products should be carefully considered when designing and optimizing Fenton-driven reme-diation systems since some of those intermediates and products are recalcitrant to further oxidation under the conditions of sub-stoichiometric Fenton reaction.

Although descriptive experimental features of the articles reviewed in this section have already been specified in the text, a comparative list of all is given in Table 5.2 together with a list of oxidation products that were identified in these studies. The list shows that phenol and p-hydroquinone were the two most com-monly observed products regardless of the applied AOP. Hydroxyacetophenone was identified in Photo-Fenton and photocatalytic processes, while methylbenzo-furan was observed in the Photo-Fenton process. The proposed reaction pathway for the Photo-Fenton process involved initiation either by direct attack of $HO^{\bullet}$ to BPA, and hydroxylation followed by abstraction of a hydrogen atom from phenolic hydroxyl groups to form quinone-like compounds, which by further oxidation ended up as aliphatic acids [87, 99, 109, 110].

Table 5.2 Summary of reaction conditions and identified oxidation intermediates for BPA removal by AOPs reviewed

Reference	Reaction conditions	Identified oxidation intermediates
Rosenfeldt and Linden [100]	BPA_o = 23.3 μM; 1 kW MPa or four 15 W LPb Hg lamps; H_2O_2 = 0–25 mg/L; UV fluence = 0–1500 mJ/cm²	
Zhou et al. [103]	BPA_o = 8.8–44 μM; 125 W HP Hg lamp (≥365 nm); pH = 3–8, [Fe(III)]:[Ox] = 10/30, 10/60, 10/120 (mM/mM)	
Katsumata et al. [99]	BPA_o = 44 μM; 990 W Xe lamp (<300 nm), I = 0.5 mW/cm²; pH= 2–4.5; Fe(II) = 0–4 × 10^{-5} M; H_2O_2 = 0–4 × 10^{-4} M	
Chen et al. [96]	BPA_o = 60 μM; 15 W LP Hg lamp (253.7 nm); H_2O_2 = 0–50 mg/L; UV fluence = 100–5,000 mJ/cm²; pH = 5.3	Phenol; 1,4-Dihydroxylbenzene; 1,4-Benzoquinone; Acetate, Oxalate
Neamtu and Frimmel [101]	BPA_o = 520 μM; 15 W LP Hg lamp (254 nm); H_2O_2 = 0–750 μM; pH = 6.7; photonic flux = 4.25 × 10^{-6} einstein s^{-1}	
Zhan et al. [102]	BPA_o = 44 μM; 500 W MP Hg vapor lamp (365 nm); I^c = 0.525 mW/cm²; humic acids = 10 mg/L	2-Hydroxy-propanoic acid; Glycerol; 4-Isopropenylphenol; p-Hydroquinone; Mono-hydroxylated BPA
Peng et al. [97]	BPA_o = 2–8 mg/L; 250 W metal halide lamp; light intensity = 150,000 lux; Chlorella vulgaris concentration = 2 × 10^9 to 12 × 10^9 cells/L; pH = 6.5 ± 0.1; humic acid = 0–6 mg/L; Fe^{3+} = 5–20 μM	
Torres et al. [107]	BPA_o = 118 μM; $FeSO_4$ = 100 μM; H_2O_2 = 35 mM; pH = 3	Monohydroxylated-4-isopropenylphenol; 4-isopropenyl phenol; 4-hydroxyacetophenone; dihydroxylated BPA; quinone of dihydroxylated BPA; monohydroxylated BPA; quinone of monohydroxylated BPA; oxalic acid; formic acid; acetic acid
Li et al. [104]	BPA_o = 0.103 mM; 8 W LZC-UV lamp with an emission peak at 365 nm; UV light intensity = 1.20 mW/cm²; iron oxide powders =0.25 g/250 mL; oxalate = 0.2–3.6 mM; pH range 2–7;	

(continued)

Table 5.2 (continued)

Reference	Reaction conditions	Identified oxidation intermediates
Ioan et al. [98]	BPA_o = 25 mg/L; $FeSO_4.7H_2O$ = 1.4 and 2.5 mg/L; H_2O_2 = 7 mg/L; pH = 4.0, 5.0 and 6.5	
Liu et al. [105]	BPA_o = 20 µM; Metal halide lamp $\lambda \geq$ 365 nm; KSF dosage = 0.5–10 g/L, pH = 3–9;	
Rodriguez et al. [106]	BPA_o = 50 µM; Parabolic collector reactor (24 L illuminated volume, 3.08 m² total irradiated surface); pH = 3, 6.5; Fe^{3+} = 50 µM; Oxalic acid = 1 mM; Citric acid = 1 mM; α-Fe_2O_3 = 100 mg/L; H_2O_2 = 1 mM	Benzaldehyde; Phenol; 1-Hexanol-2-ethyl; p-isopropenylphenol; Acetophenone, 4'-hydroxy; 4, 4'-dihydroxybenzophenone; Ethanone, 1-(4-cyclohexylphenyl)
Poerschmann et al. [108]	BPA_o = 44 µM; Fe^{2+} = 10, 20 µM; $[Fe^{2+}]_0/[BPA]_0$ = 0.23, 0.45; H_2O_2 = 100, 200 µM; $[H_2O_2]_0/[Fe^{2+}]_0$ = 10; pH = 3	monohydroxylated BPA; dihydroxylated BPA; 4-hydroxy-1-phenoxy-2-isopropyl-4-phenol; 4-isopropyl-1-phenoxy-2-isopropyl-4-phenol; 4,4-dihydroxy-methylstilbene

5.4 Concluding Remarks

Endocrine distrupting compounds (EDCs) have been demonstrated to be environmentally prevalent and concern is growing over their potential impact on human and environmental health due to their initiation of hormone-like activities even in trace concentrations. Moreover, recent studies on occurrence and fate of EDCs have shown that conventional water and wastewater treatment processes may be inadequate to fully remove these compounds from treated effluents. Alkylphenols such as nonylphenol, octylphenol and bisphenol A are the examples of EDCs which are determined throughout the most of the water and wastewater sources. From the summary provided in this chapter, it can be seen that the use of Photo-Fenton and complimentary processes are effective treatment methods for the removal of alkylphenols and bisphenol A. However, there are a number of issues to be resolved pertaining to these treatment methods. A large majority of the reviewed articles for bisphenol A involved identification of the oxidation by-products as well as intermediates however, degradation by-products are largely unknown for alkylphenols such as nonylphenol and octylphenol. Nevertheless, incomplete treatment of water and wastewater bearing these EDCs would lead to potentially toxic oxidation by-products/intermediates. Thus it is necessary to have the information on identity, properties and characteristics of the oxidation by-products/intermediates. Moreover it is also important to assess the biodegradability and potential estrogenic activity of these compounds. As a final remark, development of fast, sensitive and reliable bioassay test used for the determination of estrogenic activity provide excellent support to ensure the safety of treated water and wastewater contaminated with EDCs and to validate the treatment processes' performance as well. From the results of the laboratory studies reviewed in this chapter together with better analytical systems, more detailed and accurate models of the degradation mechanism of alkylphenols and bisphenol A, and kinetics of Photo-Fenton and complimentary processes can be developed, allowing better predictions of the behavior of these processes.

References

1. Ruppert G, Bauer R, Heisler G (1993) The Photo-Fenton reaction-an effective photochemical wastewater treatment process. J Photochem Photobiol A Chem 73(1993):75–78
2. Baxendale JH, Bridge NK (1955) The photoreduction of some ferric compounds in aqueous solution. J Phys Chem 59:783–788
3. Faust BC, Hoigné J (1990) Photolysis of iron(III)-hydroxyl complexes as sources of OH radicals in clouds, fog and rain. Atmos Environ 24A:79–89
4. Sun Y, Pignatello JJ (1993) Photochemical reactions involved in the total mineralization of 2, 4-D by $Fe^{3+}/H_2O_2/UV$. Environ Sci Technol 27:304–310
5. Zepp RG, Faust BC, Hoigné J (1992) Hydroxyl radical formation in aqueous reactions (pH 3–8) of iron (II) with hydrogen peroxide: the Photo-Fenton reaction. Environ Sci Technol 26:313–319

6. Sedlak DL, Andren AW (1991) Oxidation of chlorobenzene with Fenton's reagent. Environ Sci Technol 25:777–782
7. Kiwi J, Lopez A, Nadtochenko V (2000) Mechanism and kinetics of the OH-radical intervention during Fenton oxidation in the presence of a significant amount of radical scavenger (Cl⁻). Environ Sci Technol 34:2162–2168
8. BenkelbergH J, Warneck P (1995) Photodecomposition of iron(III) hydroxo and sulfato complexes in aqueous solution: wavelength dependence of OH and SO4− quantum yields. J Phys Chem 99:5214–5221
9. Dainton FS, Sisley WD (1963) Polymerization of methacrylamide in aqueous solution. Part 2—the ferric-ion-photosensitized reaction. Trans Faraday Soc 59:1377–1384
10. Evans MG, Uri N (1949) Photochemical polymerization in aqueous solution. Nature 164:404–405
11. De Laat J, Gallard H (1999) Catalytic decomposition of hydrogen peroxide by Fe(III) in homogeneous aqueous solution: mechanism and kinetic modeling. Environ Sci Technol 33:2726–2732
12. Balzani V, Carassiti V (1970) Photochemistry of coordination compounds. Academic Press, London, Chapter 10, pp 145–192
13. Pignatello JJ, Liu D, Huston P (1999) Evidence for an additional oxidant in the photo assisted Fenton reaction. Environ Sci Technol 33:1832–1839
14. Barbeni M, Minero C, Pelizzetti E, Borgarello E, Serpone N (1987) Chemical degradation of chlorophenols with Fenton's reagent ($Fe^{2+} + H_2O_2$). Chemosphere 16:2225–2237
15. Eisenhauer HR (1964) Oxidation of phenolic wastes. J Water Poll Control Fed 36(9): 116–1128
16. Haag WR, Yao CCD (1992) Rate constant for reaction of hydroxyl radicals with several drinking water contaminants. Environ Sci Technol 26:1005–1013
17. Murphy AP, Boegli WJ, Kevin Price M, Moody CD (1989) A Fenton-like reaction to neutralize formaldehyde waste solutions. Environ Sci Technol 23(2):166–169
18. Yoon J, Lee Y, Kim S (2004) Investigation of the reaction pathway of OH radicals produced by Fenton oxidation in the conditions of wastewater treatment. Water Sci Technol 44(5):15–21
19. Bossmann SH, Oliveros E, Göb S, Siegwart S, Dahlen EP, Payawan Jr., L, Straub M, Wörner M, Braun A (1998) New evidence against hydroxyl radicals as reactive intermediates in the thermal and photochemiclly enhanced Fenton reactions. J Phys Chem A 102:5542–5550
20. Nadtochenko VA, Kiwi J (1998) Photolysis of $FeOH^{2+}$ and $FeCl^{2+}$ in aqueous solution. Photodissociation kinetics and quantum yields. Inorg Chem 37:5233–5238
21. Buxton GV, Greenstock CL, Helman WP, Ross AB (1988) Critical review of rate constants for reactions of hydrated electrons, hydrogen atoms and hydroxyl radicals ($^{\bullet}OH/^{\bullet}O^{-}$) in aqueous solution. J Phys Chem Ref Data 17:513–886
22. Gob S, Oliveros E, Bossmann SH, Braun AM, Nascimento CAO, Guardani R (2001) Optimal experimental design and artificial neural networks applied to the photochemically enhanced Fenton reaction. Water Sci Technol 44(5):339–345
23. Lee Y, Lee C, Yoon J (2003) High temperature dependence of 2, 4-dichlorophenoxyacetic acid degradation by Fe^{3+}/H_2O_2 system. Chemosphere 51:963–971
24. Lunar L, Sicilia D, Rubio S, Perez-Bendito D, Nickel U (2000) Degradation of photographic developers by Fenton's reagent: condition optimization and kinetics for metol oxidation. Water Res 34(6):1791–1802
25. Sagawe G, Lehnard A, Lubber M, Rochendorf G, Bahnemann D (2001) The insulated solar Fenton hybrid process: fundamental investigations. Helvet Chem Acta 84(12):3742–3759
26. Solozhenko EG, Soboleva NM, Goncharuk VV (1995) Decolourization of azo dye solutions by Fenton's oxidation. Water Res 29(9):2206–2210
27. Lee C, Yoon J (2004) Temperature dependence of hydroxyl radical formation in the hf/ Fe3+/H2O2 and Fe^{3+}/H_2O_2 systems. Chemosphere 56:923–934

28. Faust BC, Zepp RG (1993) Photochemistry of aqueous iron(III)-polycarboxylate complexes: roles in the chemistry of atmospheric and surface water. Environ Sci Technol 27:2517–2522
29. Safarzadeh-Amiri A, Bolton JR, Cater SR (1997) Ferrioxalate-mediated photodegradation of organic pollutants in contaminated water. Water Res 31:787–798
30. Balmer ME, Sulzberger B (1999) Atrazine degradation in irradiated iron/oxalate systems: effects of pH and oxalate. Environ Sci Technol 33:2418–2424
31. Hatchard CG, Parker CA (1956) A new sensitive chemical actinometer. II. Potassium ferrioxalate actinometry as a standard chemical actinometer. Proc R Soc London A 253:518–536
32. Ansari A, Peral J, Domenech X, Rudeiques-Clemente R (1997) Oxidation of HSO_3^- in aqueous suspensions of alpha-Fe_2O_3, alpha-FeOOH, beta-FeOOH and gamma-FeOOH in the dark and under illumination. Environ Pollut 5(3):283–288
33. Leland JK, Bard AJ (1987) Photochemistry of colloidal semiconducting iron oxide polymorphs. J Phys Chem 91:5076–5083
34. He J, Tao X, Ma W, Zhao J (2002) Heterogenous Photo-Fenton degradation of an azo dye in aqueous H_2O_2/iron oxide dispersions at neutral pHs. Chemistry Letters, The Chemical Society of Japan, pp 86–87
35. Maletzky P, Bauer R, Lahnsteiner J, Pouresmael B (1999) Immobilization of iron ions on nafion® and it's applicability to the photo-Fenton method. Chemosphere 38:2315–2325
36. Sabhi S, Kiwi J (2001) Degradation of 2, 4-dichlorophenol by immobilized iron catalysts. Water Res 35:1994–2002
37. Maletzky P, Bauer R (1999) Immobilization of iron ions on Nafion® and its applicability to the photo-Fenton method. Chemosphere 38(10):2315–2325
38. Scott JP, Ollis DF (1995) Integration of chemical and biological oxidation processes for water treatment: review and recommendations. Environ Prog 14:88–103
39. Ballesteros Martín MM, Sánchez Pérez JA, Acién Fernández FG, Casas López JL, García-Ripoll AM, Arques A, Oller I, Malato SR (2008) Combined photo-Fenton and biological oxidation for pesticide degradation: effect of photo-treated intermediates on biodegradation kinetics. Chemosphere 70(8):1476–1483
40. Sarria V, Deront M, Péringer P, Pulgarin C (2003) Degradation of a biorecalcitrant dye precursor present in industrial wastewaters by a new integrated iron(III) photoassisted-biological treatment. C Appl Catal B 40:231–246
41. Nimrod AC, Benson WH (1996) Environmental estrogenic effects of alkylphenol ethoxylates. Crit Rev Toxicol 26(3):335–364
42. Ning B, Graham N, Zhang Y, Nakonechny M, El-Din MG (2007) Degradation of endocrine disrupting chemicals by ozone/AOPs, ozone. Environ Sci Eng 29:153–176
43. Planas C, Guadayol JM, Droguet M, Escalas A, Rivera J, Caixach J (2002) Degradation of polyethoxylated nonylphenols in a sewage treatment plant. Quantitative analysis by isotopic dilution-HRGC/MS. Water Res 36(4):982–988
44. Ying GG, Williams B, Kookana R (2002) Environmental fate of alkylphenols and alkylphenol ethoxylates—a review. Environ Int 28(3):215–226
45. Sharma VK, Anquandah GAK, Yngard RA, Kim H, Fekete J, Bouzek K, Ray AK, Golovko D (2009) Nonylphenol, octylphenol, and bisphenol-A in the aquatic environment: a review on occurrence, fate and treatment. J Environ Sci Health Part A44:423–442
46. Brook D, Crookes M, Johnson I, Mitchell R, Watts C (2005) Prioritasation of alkylphenols for environmental risk assessment. National Centre for Ecotoxicology and Hazardous Substances, Environ Agency, Bristol
47. Ahel M, Giger W, Koch M (1994) Behavior of alkylphenol polyethoxylate surfactants in the aquatic environment 1. Occurrence and transformation in river. Water Res 28:1143–1152
48. Neamtu M, Popa DM, Frimmel FH (2009) Simulated solar UV-irradiation of endocrine disrupting chemical octylphenol. J Hazard Mater 164:1561–1567

49. Oehlmann J, Schulte-Oehlmann U, Tillmann M, Markert B (2000) Effects of endocrine disruptors on prosobranch snails (Mollusca Gastropoda) in the laboratory. Part I: bisphenol A and octylphenol as xeno-estrogens. Ecotoxicology 9:383–397

50. Sores A, Guieysse B, Jefferson B, Cartmell E, Lester JN (2008) Nonylphenol in the environment: a critical review of occurrence, fate, toxicity and treatment in wastewaters. Environ Int 34:1033–1049

51. Neamtu M, Frimmel FH (2006a) Photodegradation of endocrine disrupting chemical nonylphenol by simulated solar UV-irradiation. Sci Total Environ 369:295–306

52. Solé M, de Alda MJL, Castillo M, Porte C, Ladegaard-Pedersen K, Barcelo D (2000) Estrogenicity determination in sewage treatment plants and surface waters from the Catalonian area (NE Spain). Environ Sci Technol 34:5076–5083

53. Canada (2002) Canadian environmental quality guidelines for nonylphenol and its ethoxylates (water, sediment, and soil) scientific supporting document.ecosystem health: science-based solutions report no. 1–3. National Guidelines and Standards Office, Environmental Quality Branch, Environment Canada

54. EU (2002) European union risk assessment report. 4-Nonylphenol (Branched) and Nonylphenol. 2nd priority list 10

55. Maguire RJ (1999) Review of the persistence of nonylphenol and nonylphenol ethoxylates. Water Qual Res J Can 34:37–78

56. Sonnenschein C, Soto AMJ (1998) An updated review of environmental estrogen and androgen mimics and antagonists. Steroid Biochem Mol Biol 65:143–150

57. http://www.environment-agency.gov.uk/business/topics/pollution/39131.aspx

58. Brand N, Mailhot G, Bolte M (1998) Degradation photoinduced by Fe(III): method of alkylphenol ethoxylates removal in water. Environ Sci Technol 32:2715

59. de la Fuente L, Acosta T, Babay P, Curutchet G, Candal R, Litter MI (2010) Degradation of nonylphenol ethoxylate-9 (NPE-9) by photochemical advanced oxidation technologies. Ind Eng Chem Res 49:6909–6915

60. Destaillats H, Hung HM, Hoffmann MR (2000) Degradation of alkylphenol ethoxylate surfactants in water with ultrasonic irradiation. Environ Sci Technol 34:311

61. BKH (2000) EUROPEAN COMMISSION DG ENV Towards the establishment of a priority list of substances for further evaluation of their role in endocrine disruption—preparation of a candidate list of substances as a basis for priority setting, Final Report, BKH consulting engineers, Delft, The Netherlands in association with TNO nutrition and food research, Zeist, The Netherlands

62. USEPA (2005) Ambient aquatic life water quality criteria-Nonylphenol Final. Office of Water, Office of Science and Technology, Washington, DC. EPA-822-R-05-005

63. OSPAR Commission (2003) Hazardous Substances Series Octylphenol

64. Butwell AJ, Hetheridge M, James HA, Johnson AC, Young WF (2002) Endocrine disrupting chemicals in wastewater: a review of occurrence and removal. UK Water Industry Research Limited, London

65. Rojas MR, Pérez F, Whitley D, Arnold RG, Sáez AE (2010) Modeling of advanced oxidation of trace organic contaminants by hydrogen peroxide photolysis and Fenton's reaction. Ind Eng Chem Res 49:11331–11343

66. Chen PJ, Rosenfeldt EJ, Kullman SW, Hinton DE, Linden KG (2007) Biological assessments of a mixture of endocrine disruptors at environmentally relevant concentrations in water following UV/H_2O_2 oxidation. Sci Total Environ 376:18–26

67. Litter MI (2005) Introduction to photochemical advanced oxidation processes for water treatment. In: Boule P, Bahnemann DW, Robertson PKJ, (eds) The handbook of environmental chemistry, vol. 2, Part M, pp 325–366, Springer, Berlin

68. Oppenländer T (2003) Photochemical purification of water and air advanced oxidation processes (AOPs): principles, reaction mechanisms, reactor concepts. Wiley, New York

69. Pera-Titus M, García-Molina V, Baños MA, Giménez J, Espulgas S (2004) Degradation of chlorophenols by means of advanced oxidation processes: a general review. Appl Catal B Environ 47:219

70. Arslan-Alaton I, Shayin S, Olmez-Hanci T (2011) The Hydroxyl radical scavenging effect of textile preparation auxiliaries on the photochemical treatment of nonylphenol ethoxylate. Environ Technol (in press)

71. Kim J, Korshin GV, Velichenko AB (2005) Comparative study of electrochemical degradation and ozonation of nonylphenol. Water Res 39:2527–2534

72. Mizuno T, Yamada H, Tsuno H (2002) Characteristics of oxidation by-products formation during ozonation and ozone/hydrogen peroxide process in the aqueous solution of nonylhenol ethoxylates. Adv Asian Environ Eng 2(2):33–42

73. Sherrard KB, Marriott PJ, Amiet RG, McCormick MJ, Colton R, Millington K (1996) Spectroscopic analysis of heterogeneous photocatalysis products of nonylphenol- and primary alcohol ethoxylate nonionic surfactants. Chemosphere 33(10):1921–1940

74. http://www.ewg.org/chemindex/chemicals/bisphenolA

75. Fromme H, Küchler T, Otto T, Pilz K, Müller J, Wenzel A (2002) Occurrence of phthalates and bisphenol A and F in the environment. Water Res 36:1429–1438

76. Mannsville (2008a) Chemical products synopsis: bisphenol A. Mannsville Chemical Products Corp

77. Staples CA, Dorn PB, Klecka GM, O'Block ST, Harris LR (1998) A review of the environmental fate, effects, and exposures of bisphenol A. Chemosphere 36:2149–2173

78. Crathorne B, Palmer CP, Stanley JA (1986) High-Performance liquid-chromatographic determination of bisphenol a diglycidyl ether and bisphenol-f diglycidyl ether in water. J Chromatogr 360(1):266–270

79. USEPA (2010) Bisphenol A action plan, (CASRN 80-05-7), [CA Index Name: Phenol, 4,4'-(1-methylethylidene)bis-]. http://www.epa.gov/oppt/existingchemicals/pubs/actionplans/bpa_action_plan.pdf

80. Cousins IT, Staples CA, Klecka GM, Mackay D (2002) A multimedia assessment of the environmental fate of bisphenol A. HERA 8:1107–1135

81. Japan Environment Agency (2001) Survey of endocrine disrupting substances (environmental hormones) in the aquatic environment (FY2000), available on the Internet at http://www.nies.go.jp/edc/edcdb/HomePage_e/medb/MEDB.html

82. Yamamoto T, Yasuhara A, Shiraishi H, Nakasugi O (2001) Bisphenol A in hazardous waste landfill leachates. Chemosphere 42(4):415–418

83. Bisphenol A Global Industry Group (2002) Bisphenol A information sheet

84. West RJ, Goodwin PA, Klecka GM (2001) Assessment of the ready biodegradability of bisphenol A. Bull Environ Contam Toxicol 67:106–112

85. Dorn PB, O'Block ST, Harris LR (1998) A review of the environmental fate, effects, and exposures of bisphenol A. Chemosphere 36:2149–2173

86. Klecka GM, Gonsior SJ, West RJ, Goodwin PA, Markham DA (2001) Biodegradation of Bisphenol A in aquatic environments: river die-away. Environ Toxicol Chem 20:2725–2735

87. Gültekin I, Ince NH (2007) Synthetic endocrine disruptors in the environment and water remediation by advanced oxidation processes. J Environ Manage 85:816–832

88. Jintelmann J, Katayama A, Kurihara N, Shore L, Wenzel A (2003) Endocrine disruptors in the environment. Pure Appl Chem 75:631–681

89. Crain DA, Eriksen M, Iguchi T, Jobling S, Laufer H, LeBlanc GH, Guillette LJ (2007) An ecological assessment of bisphenol-A: evidence from comparative biology. Reprod Toxicol 14:225–239

90. Canada (2008) Screening assessment for the challenge phenol, 4,4' (1-methylethylidene)bis-(Bisphenol A) CAS 80-05-7, Environment Canada http://www.ec.gc.ca/substances/ese/eng/challenge/batch2/batch2_80-05-7_en.pdf

91. USEPA (2009) Drinking water contaminant candidate list and regulatory determinations, contaminant candidate list 3. http://www.epa.gov/ogwdw000/ccl/ccl3.html

92. AIST (2007) (Japan's National Institute of Advanced Industrial Science and Technology) AIST risk assessment document series 4. Bisphenol A

93. EU (2008) European Union updated risk assessment report. Bisphenol A, CAS No: 80-05-7. Institute for Health and Consumer Protection, European Chemicals Bureau, European Commission Joint Research Centre, 3rd Priority List, Luxembourg

94. European Food and Safety Authority (EFSA) (2008) Scientific opinion of the panel on food additives, flavourings, processing aids and materials in contact with food (AFC) on a request from the commission on the toxicokinetics of bisphenol A. EFSA J 759:1–10

95. Mohapatra DP, Brar SK, Tyagi RD, Surampalli RY (2010) Physico-chemical pre-treatment and biotransformation of wastewater and wastewater sludge-fate of bisphenol A. Chemosphere 78:923–941

96. Chen PJ, Linden KG, Hinton DE, Kashiwada S, Rosenfeldt EJ, Kullman SW (2006) Biological assessment of bisphenol A degradation in water following direct photolysis and UV advanced oxidation. Chemosphere 65:1094–1102

97. Peng Z, Wu F, Deng N (2006) Photodegradation of bisphenol A in simulated lake water containing algae, humic acid and ferric ions. Environ Pollut 144:840–846

98. Ioan I, Wilson S, Lundanes E, Neculai A (2007) Comparison of Fenton and sono-Fenton bisphenol A degradation. Hazard Mater 142:559–563

99. Katsumata H, Kawabe S, Kaneco S, Suzuki T, Ohta K (2004) Degradation of bisphenol A in water by the photo-Fenton reaction. J Photochem Photobiol A 162:297–305

100. Rosenfeldt EJ, Linden KG (2004) Degradation of endocrine disrupting chemicals bisphenol A, ethinyl estradiol, and estradiol during UV photolysis and advanced oxidation processes. Environ Sci Technol 38:5476–5484

101. Neamtu M, Frimmel FH (2006b) Degradation of endocrine disrupting bisphenol A by 254 nm irradiation in different water matrices and effect on yeast cells. Water Res 40(20):3745–3750

102. Zhan M, Yang X, Xian Q, Kong L (2006) Photosensitized degradation of bisphenol A involving reactive oxygen species in the presence of humic substances. Chemosphere 63:378–386

103. Zhou D, Wu F, Deng N, Xiang W (2004) Photooxidation of bisphenol A (BPA) in water in the presence of ferric and carboxylate salts. Water Res 38:4107–4116

104. Li FB, Li XZ, Liu CS, Li XM, Liu TX (2007) Effect of oxalate on photodegradation of Bisphenol A at the interface of different iron oxides. Ind Eng Chem Res 46(3):781–787

105. Liu YX, Zhang X, Guo L, Wu F, Deng NS (2008) Photodegradation of Bisphenol A in the montmorillonite KSF suspended solutions. Ind Eng Chem Res 47(19):7141–7146

106. Rodríguez EM, Fernández G, Klamerth N, Maldonado MI, Álvarez PM, Malato S (2010) Efficiency of different solar advanced oxidation processes on the oxidation of bisphenol A in water. Appl Catal B 95:228–237

107. Torres RA, Abdelmalek F, Combet E, Pétrier C, Pulgarin C (2007) A comparative study of ultrasonic cavitation and Fenton's reagent for bisphenol A degradation in deionised and natural waters. J Hazard Mater 146:546–551

108. Poerschmann J, Trommler U, Górecki T (2010) Aromatic intermediate formation during oxidative degradation of bisphenol A by homogeneous sub-stoichiometric Fenton reaction. Chemosphere 79:975–986

109. Fukahori S, Ichiura H, Kitaoka T, Tanaka H (2003) Capturing of bisphenol A photodecomposition intermediates by composite TiO2-zeolite sheets. Appl Catal B 46:453–462

110. Kaneco S, Rahman MA, Suzuki T, Katsumata H, Ohta K (2004) Optimization of solar photocatalytic degradation conditions of bisphenol A in water using titanium dioxide. J Photochem Photobiol A 163:419–424

Chapter 6
Outlook

Giusy Lofrano

At the end of this trip among technologies and pollutants, we may figure out to draw a relationship linking environment, green engineering, green chemistry and sustainability, as shown in Fig. 6.1, where springs sets permit to connect ideally the different disciplines in a single system.

There is no one technology that solves all issues under all conditions. As an example, significant environmental properties of MBR technology should be evaluated in relation to increased environmental costs mainly due to membrane and equipment production and additional energy consumption, although the effluent of MBR plants can exert lower estrogenic activity, in comparison with conventional activated sludge.

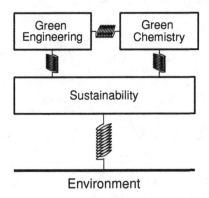

Fig. 6.1 Relationships among environment, sustainability, green engineering and green chemistry

G. Lofrano (✉)
Department of Civil Engineering, University of Salerno,
via ponte don Melillo, 84084 Fisciano (SA), Italy
e-mail: glofrano@unisa.it

G. Lofrano (ed.), *Green Technologies for Wastewater Treatment*,
SpringerBriefs in Green Chemistry for Sustainability,
DOI: 10.1007/978-94-007-1430-4_6, © Lofrano 2012

On the other hand, all technologies investigated so far have shown a high capacity for removing organics, offering a series of environmental advantages which may suggest them to be promising green technologies. Further research remains to be accomplished before some of those advantages become established technology for water and wastewater treatment.